D0403101

Shadow OF A Star

Shadow

OF A Star

THE NEUTRINO
STORY OF
SUPERNOVA 1987A

ALFRED K. MANN

 W.H. Freeman and Company
NEW YORK

Cover Designer: Anne Scatto/Pixel Press
Text Designer: Blake Logan

Library of Congress Cataloging-in-Publication Data

Mann, Alfred K.
    Shadow of a star : the neutrino story of Supernova
1987A / Alfred K. Mann.
        p.   cm.
    Includes bibliographical references and index.
    ISBN 0-7167-3097-9
    1. Supernova 1987A.   2. Neutrino astrophysics.
    I. Title.
QB843.S95M26   1997
523.8'4465—dc21   96-29967
                                                        CIP

Published in the United States of America

First printing, 1997

To my wife, Jayne, who has
countenanced my love affair
with physics for these fifty years

CONTENTS

PREFACE

A personal comment on the mostly impersonal tone of this book is probably in order here. The work I describe in this book was done by many people of different ages and backgrounds: graduate students, postdoctoral research associates, physicist-engineers, and professors. The occasional inclusion of an anecdote about a few people identified by name makes for easier reading and, I hope, for closer contact between the subject matter and the reader. But this is principally the story of a star and not of the humans who studied it. Consequently, there may be fewer personal reminiscences and references to individual people by name than some might like.

Nevertheless, credit should be given where due, and accordingly the first illustration in the book consists of copies of the documents awarding the Bruno Rossi Prize of the High Energy

Astrophysics Division of the American Astronomical Society to the individuals, appropriately recognized by name, who carried out the work of science described here.

In addition are the personal acknowledgments that I am glad to make. The leader in name, experience, and talent of the University of Tokyo physicists was Professor Masatoshi Koshiba, whose foresight and good physics sense were responsible for the construction of the massive apparatus in Kamioka, Japan, which has a central part in this story. It was at Professor Koshiba's invitation that our scientific collaboration was formed, and it was at his insistence that the collaborative effort continued in a spirit of warm cooperation, holding to the important physics goals that it ultimately achieved. He and I, similar in age, were essentially strangers when our collaboration began, but in him I found a kindred soul for whom my respect and affection have increased over time. I also enjoyed and profited from the intellectual and emotional exchanges with my friends and colleagues Teruhiro Suda, Yoji Totsuka, and Atsuto Suzuki. Unfortunately, Suda died prematurely, but Totsuka and Suzuki are now firmly established as professors themselves. And the Japanese students, as are students everywhere, were a tower of strength.

In my own institution, the University of Pennsylvania, I have worked for many years in many experiments with Professor Eugene W. Beier, who shared with me the direction of the Penn effort at Kamioka. The Penn students—Weiping Zhang, Soo Bong Kim, and Edward Frank, with Professor William Frati, who joined us somewhat later— were a joy to work with and to live with side by side in a country so different from our own.

I am grateful to Jean O'Boyle, who typed and corrected the manuscript until it satisfied us both and to Joseph Guerrero who was responsible for the figures. In Chapter 12, the photographs and the material relating to the Hubble Space Telescope and its findings were provided by Dr. Jeffrey J. E. Hayes of the Space Telescope Science Institute and Dr. Christopher Burrows of the European Space Agency/Space Telescope Science Institute.

I am especially grateful to those who read and reviewed the manuscript: Gerald E. Brown, Joseph F. Dolan, Georg G. Raffelt, Dennis W. Sciama, and Joseph I. Silk. They made many constructive suggestions of style and substance that significantly improved the book.

Finally, Holly Hodder, the editor, tended this book as if it were a garden—one moment gently nurturing, the next viciously pruning, and always weeding. I owe her my heartfelt thanks.

The High Energy Astrophysics Division

of the

American Astronomical Society

hereby awards the

Bruno Rossi Prize for 1989

to the members of the

Kamiokande and IMB

high energy neutrino collaborations.

The Prize is awarded for the dramatic and mutually confirming detection of a burst of neutrinos from supernova 1987A, which opened a new window on the cosmos beyond the solar system and provided the first direct data on the high-energy processes occurring in the centers of collapsed stars. The names of the members of the Kamiokande Collaboration are given below:

K. Hirata	T. Tanimori
T. Kajita	K. Miyano
M. Koshiba	M. Yamada
M. Nakahata	E. W. Beier
Y. Oyama	L. R. Feldscher
N. Sato	S. B. Kim
A. Suzuki	A. K. Mann
M. Takita	F. M. Newcomer
Y. Totsuka	R. Van Berg
T. Kifune	W. Zhang
T. Suda	B. Cortez
K. Takahashi	

John Wheeler
Chairman, H.E.A.D.
3 October 1989
Date

Robert C. Haymes
Secretary-Treasurer, H.E.A.D.
October 1989
Date

The High Energy Astrophysics Division
of the
American Astronomical Society
hereby awards the
Bruno Rossi Prize for 1989
to the members of the
IMB and Kamiokande
high energy neutrino collaborations.

The Prize is awarded for the dramatic and mutually confirming detection of a burst of neutrinos from supernova 1987A, which opened a new window on the cosmos beyond the solar system and provided the first direct data on the high-energy processes occurring in the centers of collapsed stars. The names of the members of the IMB Collaboration are given below:

Richard M. Bionta	Kenneth S. Ganezer	Frederick Reines
Geoffrey Blewitt	Maurice Goldhaber	Jonas Schultz
Clyde B. Bratton	Todd J. Haines	Sally C. Seidel
David W. Casper	Tegid W. Jones	Eric L. Shumard
Alessandra Ciocio	Danuta Kielczewska	Daniel A. Sinclair
Richard Claus	William R. Kropp	Henry W. Sobel
Bruce Cortez	John G. Learned	James L. Stone
Marshall Crouch	John M. LoSecco	Lawrence R. Sulak
Stephen T. Dye	James M. Matthews	Robert R. Svoboda
Steven M. Errede	Richard Miller	Gregory J. Thornton
G. William Foster	Manjeet S. Mudan	John C. van der Velde
Wojciech Gajewski	Hye-Sook Park	Craig R. Wuest
	LeRoy Price	

J. Craig Wheeler
Chairman, H.E.A.D.

31 October 1989
Date

Robert C. Haymes
Secretary-Treasurer, H.E.A.D.

27 October 1989
Date

1 THE WHY AND WHEREFORE OF THIS BOOK

In a series of lectures in 1965 at the Brookhaven National Laboratory in New York, Barbara Ward, a noted British economist, began with the following paragraph:

> In the last few decades, mankind has been overcome by the most fateful change in its entire history. Modern science and technology have created so close a network of communication, transport, economic interdependence—and potential nuclear destruction—that planet earth, on its journey through infinity, has acquired the intimacy, the fellowship, and the vulnerability of a spaceship.

Professor Ward went on to discuss the implications of that "most fateful change" and to make some suggestions that might help us adapt to it. In the three decades since she gave those lectures, society has struggled to implement some of her ideas. At the same time, however, the rate of change has accelerated, thereby compounding the problem.

Fueled by the industrial and computer revolutions and the concomitant steadily rising standard of living in the industrialized nations, our understanding of the breadth and depth of the universe has expanded unceasingly in the twentieth century. In the physical sciences, our range now extends from distances a thousand times smaller than the radius of the atomic nucleus to distances enormously greater than the radius of the Sun's planetary system. Underlying these developments are elegant mathematical formulations—still far from complete but with astounding predictive powers. Perhaps more remarkable has been the progress in the life sciences, in which the deciphering of the genetic code has given new vitality and a vastly extended purview to biology and its related disciplines. Similar statements may be made about the progress in chemistry, in the science of materials, in the earth sciences, and in psychology.

This scientific progress has in its turn provided new impetus to the technological revolutions of our era and has contributed to the relative ease with which many of us live today. A case in point is the quality of medical treatment in the scientifically advanced nations. In these countries, medicine has been transformed from an art with a scientific underpinning to a powerful, predictive science, the practice of which is an art. This transformation has come about as experts in the basic sciences—frequently with a Ph.D as well as an M.D.—have brought to medicine the advances in research in their respective scientific fields. Clinicians, always pragmatic and resourceful, seized these advances to ease pain and suffering and lengthen life expectancy.

Despite these accomplishments, however, society still regards the pursuit of fundamental science warily and with mixed feelings. One reason is the widespread deep feeling of insecurity engendered by our inability to adapt to the rapid, extensive "fateful change" referred to earlier. A related reason is that a major step forward in science, unlike a major accomplishment in art, is hard for nonscientists to appreciate, except possibly in the beneficial, practical application of it. Still another reason, and not a trivial one by any means, is the commonly held but fallacious thought that pure science is a less passionate activity than art. Many people find it difficult to empathize with the practitioners of the seemingly less human, more cold-blooded pursuit.

It is correct that a work of science is not like a work of art. To be moved by a work of art, to understand it at some level, does not require us to be particularly knowledgeable or versed in the scholarly assessment of it. That is, a work of art stands alone and may be savored for its individual worth. On the other hand, a work of science rarely stands alone, and so before we can appreciate it, we need some familiarity with its antecedents and with the nearby areas of science on which it casts new light. A good illustration can be found in the closing sentence of the famous, brief paper by James Watson and Francis Crick explaining the structure of DNA: "It has not escaped our notice that the specific pairing we have postulated immediately suggests a possible copying mechanism for the genetic material." In that sentence, perhaps because of its understatement, the natural extension of their important discovery to other areas of science is clearly outlined, even for those only marginally literate in biology.

We know now that the "specific pairing" referred to in this sentence provides the mechanism by which living cells replicate themselves. The redundancy of the double-stranded DNA helix makes possible the mechanical and chemical stability of DNA necessary for its survival in the sometimes harsh world of the living organism and enables the genetic constitution of each cell to be copied. Of course, we were not aware of all of this at the time the original work was done—and certainly not of the universal nature of DNA that has since been recognized—but all who had even a passing acquaintance with the biology of that period and an elementary knowledge of the methods of replication offered by templates and computer iterations could see, if only dimly, that Watson and Crick had done a great work of science.

The challenge facing scientists is to find ways to convey the essential features, elegance, and simplicity of important works of science so that those features may be appreciated, just as they are appreciated in important works of art. We need a discipline that might be called "science appreciation" to go along with the well-established discipline of art appreciation. There are practical advantages to scientists in helping develop such appreciation. For most scientists, however, the desire to have their work better understood is motivated principally by the idea that the culture of science—which they regard as a precious part of the modern age—should be shared with and understood by the society that is immersed in it and yet apart from it. In this they have the same motivation as the artists.

Among the many attempts to share and transmit this understanding, the late Lewis Thomas's essays on biology and medicine stand out for their clarity, originality, and humility. Similarly, Stephen Jay Gould's discussions of the origin and extinction of species and related subjects have been influential in explaining the culture of one area of science to scientists and nonscientists alike. It has been somewhat more difficult to achieve the same kind of exposition in the physical sciences, because the technical language is different and the repository of the lore is in equation form. But books by Carl Sagan, Steven Weinberg, and Jeremy Bernstein, for example, have been successful in describing to nonspecialists the grandeur of present-day thinking in astronomy, cosmology, and particle physics.

In addition to those overviews of broad areas of science, there is also a need to present single works of science as significant entities in their own right, to outline their motivation and development (akin to the sketches preceding a painting or a concerto), and to explain their meaning. It is these single works of science, supplemented by wider-ranging unifying studies, that might form the basic material of science appreciation. Not every work of science fits that mold or lends itself to that kind of individual treatment, but many—including the subject of this book—do. My story describes the detection of neutrinos carrying news of the death of a star in another galaxy, how that observation was made, and the scientific explanation of the event itself. It is a dramatic story in which the protagonist is a dying star, a supernova, whose last convulsive gasps were witnessed and recognized as part of the incessant flow of birth and death.

If my storytelling is successful, the reader may come away with an appreciation of the life cycle of stars and of the role of natural law in orchestrating the final moments of that cycle. In this story, the related sciences of atomic and nuclear physics and elementary particle physics dominate the action in the interior of the star from early in its life to its end many millions of years later. Here we meet the constituents of the nucleus—protons and neutrons—and the elementary particles—photons, electrons, and neutrinos—all actors in the unfolding drama of the supernova. And here we glimpse the intrinsic unity of the supernova phenomenon or, if you

7

prefer, the naturalness and inevitability of it. It is an event with far more violence than humans can produce, but unlike much of our violence, it is not mindless.

The next chapter summarizes the exciting first visual observations of the supernova, serving as a brief, factual description of the event similar to the front-page account of a newsworthy item in a daily newspaper. We need to do more than that, however, to achieve our purpose; we need to supply the equivalent of the analyses and editorials that a newspaper would carry as the implications of the news item unfold. Accordingly, Chapters 3 and 4 introduce the other, seemingly peripheral characters in the story, who emerge later as central to the action. When we return to the supernova itself in later chapters, you will see the event from a perspective gained from familiarity with the other characters in the story.

It would be a bonus if nonmathematical readers also grasped the enormous parameters of the supernova story. Although we present most of the numerical quantities as they usually are encountered in daily life, we must necessarily give some—the very large and the very small—in the notation used in physics and astronomy. To give these quantities meaning without disrupting the story and to help with two other issues that must be explained more fully, we have included three short appendices and a glossary at the end of the book. A short bibliography also is available for those who want to expand their horizons beyond the confines of this book.

Finally, I should note that the tenth anniversary of the observation of the supernova of this story is approaching. I hope that the story will appeal to you for its excitement and dramatic impact just as it did to scientists and nonscientists when the event occurred and as it still does to many of us.

2 A NEW SIGHTING

In the early morning of February 24, 1987, while keeping a night watch at the Las Campanas Observatory in northern Chile, Ian Shelton left the telescope building to check by eye on something strange in one of his recently taken photographs. As he scanned the sky, Shelton, a resident observer from the University of Toronto in Canada, confirmed the presence of the extraordinarily bright object in his photograph, an object that had not been in the sky earlier that night. To give you an idea of what Shelton saw, the photographs in Color Plate 1 show the same portion of the sky, with the right side taken well before February 24 and the left side a few weeks later. The difference is unmistakable. Shelton, with others, soon reported the observation of a "fifth-magnitude object, ostensibly a supernova" to the International Astronomical Union, which quickly alerted astronomers around the world to this new sighting.

Observers on sky watches often take several photographs of the same region of the sky on the same or successive nights, for reference purposes. This was what Robert McNaught, working in New South Wales, Australia, was doing on the night of February 23. McNaught photographed the same region of the sky as Shelton did and even developed the plates. Taken at times as much as 16 hours earlier than Shelton's photographs, McNaught's plates showed the supernova clearly. But he was busy with other chores and so did not then take the time to inspect them. As a result, he missed being the first person to see with the naked eye and report the presence of a new, remarkably bright object.

By midmorning on February 24, scientists throughout the world were aware of the new sighting. If indeed it was a supernova close enough to be seen with the naked eye, it was an event that had not occurred in any of their lifetimes or in the lifetime of anyone in the previous 383 years.

Optical and radio telescopes in the Southern Hemisphere were pointing at this new discovery or were preparing to do so. In the next few days, the bright object was confirmed as a supernova, the result of the explosion of one of the stars in a galaxy known as the Large Magellanic Cloud (LMC), a satellite of the Milky Way. After several more weeks of hard study, it became clear that the star that had exploded was well known as star number 202, in a group of stars 69° south of the equator, and was named Sanduleak 69° 202 (or Sanduleak 202) after Nick Sanduleak of Case Western Reserve University, who had prepared an LMC catalog in 1969. The supernova in turn entered the astronomical literature as SN 1987A, SN for supernova and A for the first supernova of the year 1987.

In the few seconds required for its destruction, Sanduleak 202 released more energy than all the energy from all the stars in the Milky Way in one year, and it decorated the southern sky with the bright light seen in Color Plate 1. Supernovae occur frequently in the many galaxies populating the sky. But a supernova so close to Earth as to be clearly visible to the naked eye is a rare happening—the last recorded one was in 1604—and consequently SN 1987A immediately caused enormous excitement.

The proximity to Earth of SN 1987A, which made it visible to the eye, also made possible the observation of a type of emission never before seen coming from an exploding star. From this we were able to learn precisely what occurred in the interior of Sanduleak 202 and what ultimately happened to it. How we found out and our resultant understanding of the explosion of Sanduleak 202 and its aftermath make up the story in this book.

The story of SN 1987A has two themes. The first is the high drama of the enormity of the event and its influence on our view of the universe, and the second is the wish, the need, prompted by human curiosity, to understand the event from a factual, rational point of view. The first theme flows easily and directly from the supernova phenomenon itself. But the second theme requires that we understand the hierarchy of the elementary particles in the nuclear world and, in particular, one inhabitant of that world— the little-known neutrino—which played a vital role in the story of SN 1987A. It was this submicroscopic world with its rigorous rules of behavior that determined the fate of SN 1987A. To this end, we shall spend the next two chapters describing two particles, the photon and the neutrino, which are basic to our understanding of what befell SN 1987A. After this brief digression, the supernova again emerges as the leading character in the story and dominates it for the remainder of this book.

3 LIGHT, NUCLEAR FUSION, AND NEUTRINOS

From earliest times, humans have studied the sky with the naked eye and remarkable skill. By the late sixteenth century, for example, the Danish astronomer Tycho Brahe had made exact observations of the planets, from which Johannes Kepler—originally Brahe's assistant and later a professor of astronomy in Germany—established the laws of planetary motion that describe our solar system. Soon thereafter, having heard of a simple magnifying instrument in Holland, the great Galileo Galilei in Padua, Italy, constructed the first complete astronomical telescope.

Since that time in the early seventeenth century, the main source of information about the heavens has been visible light observed with optical telescopes, similar to the one in Color Plate 2. But visible light is not the only way in which stars are now observed. Stars also emit waves in different amounts and at different frequencies than those of visible light, all of which constitute the family of electromagnetic radiation. For example, many stars emit waves in the radio band—the same band of frequencies on which we listen to the weather report—and some stars also emit waves at infrared and ultraviolet frequencies. The telescopes for observing these emissions have gained greater sensitivity and higher resolution as modern technology has advanced. One example is the high-resolution array of radio telescopes located in Socorro, New Mexico, which has yielded important observations of the central region of the Milky Way. A photograph of the array, capped by a rainbow, is shown in Color Plate 3, as much to display its elegant beauty as to contrast it with the optical telescope in Color Plate 2 and to indicate the magnitude of the effort required to build it. Our sense of achievement is fleeting, however, and Socorro's telescopes may be regarded in the distant future as the Stonehenge of the twentieth century, prompting scientists to ask how our primitive civilization could have known enough to create them.

Despite our 100 years of experience with it, electromagnetic radiation still is difficult to describe in nonscientific terms. But we try. First, electromagnetic radiation is a wave, and like the more familiar oceanic waves, it has crests and troughs whose rate of repetition is specified by a frequency. For example, we speak of the alternating electrical current that powers our homes as having a frequency of 60 cycles per second, a cycle being measured from, say, crest to crest of the wave. Sound waves also are described by a frequency; for instance, we refer to the low-frequency and high-frequency responses of audio equipment. What constitutes the crests and troughs of electromagnetic waves is less obvious than in the case of electrical current or sound. Here, crests and troughs are a combination of electric and magnetic fields oscillating, that is, changing repetitively in time and space, as the wave itself moves along its path. The term *field* is used in physics more or less as it is in common usage: to denote a spatial region of activity. In our case, field refers to a region of combined electric and magnetic activity traveling with the wave. These electromagnetic waves and their constituent fields do not need a medium in which to travel: no wire as the medium of electrical current; no gas, liquid, or solid as the medium of sound. Instead, these waves are launched into space from a source—for example, a flashlight or the tower of a radio station—and travel until intercepted by a gas, a liquid, or a solid, in which they deposit their electric and magnetic energy, after which they are no more.

In the twentieth century, we discovered that the amount of energy carried by a single ray of electromagnetic radiation is directly proportional to the magnitude of its frequency. Strictly speaking, however, this statement is correct only for a single ray (soon to receive its proper scientific name) and not for a wave made up of many rays, but the distinction is not important for the main point we want to make here, and so we shall temporarily use the more familiar word *wave*. The higher the frequency is, the more energy the electromagnetic wave will carry, and conversely, the lower the frequency is, the less energy the wave will carry. Note that this is not an intuitively obvious idea or even an easily verified fact outside an atomic physics laboratory. Indeed, recognizing this phenomenon required the insights of particularly gifted physicists and earned Nobel Prizes for Max Planck in 1918 and Albert Einstein in 1921.

Nevertheless, this phenomenon is a fact, and it means that among other things, the energy carried by an ultraviolet light wave is greater than the energy carried by a visible light wave. The reason is that the fields of the ultraviolet wave oscillate at a higher frequency than do those of a visible wave, which explains why humans are tanned by the ultraviolet rays from the Sun and not by the visible rays from a desk lamp. It means also that the energy carried by X rays—another wave in the family of electromagnetic radiation, differing from visible and ultraviolet radiation only by a range of frequencies higher than theirs—is greater than that carried by ultraviolet rays and consequently is more penetrating than the energy of ultraviolet rays. It is the reason that overexposure to X rays is damaging to humans and that as a consequence X-ray technicians operate their equipment from behind protective walls. Gamma rays, the historical name given to all electromagnetic radiation with frequencies appreciably higher than that of X rays—for example, gamma rays from the spontaneous decay of radioactive nuclei—are still higher in energy and still more penetrating; hence our dread of the radioactivity left after the destruction of a nuclear reactor or a nuclear weapon. All, however, are part of the family of electromagnetic radiation, with the same oscillating electric and magnetic fields. Only the frequency of oscillation is different among the family members.

The idea that the energy of a light wave is proportional to its frequency emerged well before the formulation of the dominant theory we now know as *quantum theory*. It emerged most clearly in the phenomenon of the photoelectric effect, which we use daily in the form of photoelectric cells that, for example, signal our passage and open doors automatically for us in supermarkets and hospitals. It was the work of Planck and Einstein that provided the quantitative explanation of the photoelectric effect. But their explanation was not in terms of light as waves but of light as particles, another seemingly contradictory aspect of light.

Today, we recognize the dual nature of electromagnetic radiation as implicit in quantum theory. That is, light displays a wave nature in certain phenomena and a particle nature in other phenomena. In other words, we describe certain observed phenomena most simply and accurately by treating light as if it is a wave, whereas the light in other phenomena consists of particles, in the usual meaning of that word. These two facets of the nature of light are distinct but inextricably related in the quantum theory. Indeed, in any phenomenon, light can be regarded in the quantum theory as either wave or particle as long as the remainder of the physical system involved with the light is interpreted in a consistent way. The energy of the particle of light is therefore given by a numerical constant (named for Planck) multiplied by the frequency of the "associated" wave, that is, the wave implicit in the dual nature of light. In modern physics, we speak of waves of light and particles of light. The particles we call *photons*, after their photoelectric and photochemical effects. Photons should not be confused with the *proton*, the nuclear particle, despite the similarity of their names.

Besides electromagnetic radiation, another source of vital information about stars is a product of energetic nuclear reactions in a star and is fundamentally different from electromagnetic radiation in origin and nature. We have known for many years that nuclear fusion is what generates energy in stars. This is a process in the core of the star, involving several nuclear reactions that ultimately result in four hydrogen atoms' (actually hydrogen nuclei, that is, protons) becoming transmuted and fused into a helium atom while simultaneously releasing a large quantity of energy. One part of this energy indirectly gives rise to the light of many frequencies emitted from the surface of the star, and another part carries the information that directly describes the events at the star's center.

From small-scale studies of nuclear reactions in the laboratory, the fusion process can be reproduced well enough to provide a detailed understanding of energy generation in a star such as our Sun. From these studies we also learned that some of the energy released in the fusion process is not in the form of light, but of other equally remarkable elementary particles known as *neutrinos*.

Neutrinos are not members of the family of electromagnetic radiation; they are not created or absorbed in electric or magnetic phenomena; and they do not carry electric and magnetic fields with them. The existence of neutrinos was first proposed at the end of 1930 by Wolfgang Pauli in Switzerland, slightly more than a year before James Chadwick discovered the other nuclear particle, the *neutron*. The neutron and the proton are the components of atomic nuclei; the mass and electric charge of all nuclei are combinations of protons and neutrons held together by the *strong*, or *nuclear*, *force*, one of the four fundamental forces in nature. Once it was realized that the neutrino and the neutron had to be electrically neutral and that the neutrino had a very small mass relative to that of the neutron, it was natural to give it the name it continues to bear. Neutrinos do not react to electrical charges on other matter, nor do they have any significant gravitational attraction to other matter on any but cosmic scales. In the laboratory, neutrinos are produced in the spontaneous disintegrations of radioactive nuclei and elementary particles and may disappear through the inverse processes, that is, by being captured by a nucleus or an elementary particle. Neutrinos belong to a general class of particles exerting a force on one another that is so intrinsically weak that it is often referred to as the *weak*, or *Fermi*, *force*, named for the Italian physicist Enrico Fermi, who quantitatively described the neutrinos' properties and was responsible for their name.

It is important to understand the meaning of the adjective *weak* as used in reference to neutrinos. A simple thought experiment may be useful. Think of a number of particles, in this instance neutrinos, with the same energy, all moving in the same direction. This is called a *beam*. Suppose there is a fixed number of neutrinos in the beam and we wish to remove one-half of them from the beam by putting into it a given thickness of an absorbing material and leaving the other half in the beam downstream of the absorber. What thickness of absorber do you think might be required to accomplish this?

If we were talking about a beam of visible light, you might suggest an absorber of thin paper or wood such as a venetian blind. If we were talking about X rays, we would need a lead shield similar to the one the dentist uses to cover you when X-raying your teeth. In these instances, a substantial number of photons corresponding to the rays of visible light or X rays are removed. But you would find that to absorb one-half the neutrinos from the beam, even the entire thickness of the Earth would not be enough, nor would 10 times that thickness. Roughly speaking, removing one-half the neutrinos would require an absorber (of any material you choose) with a thickness equal to the entire distance between the Earth and the Sun. That is, the interaction of neutrinos with matter is so weak as to make all matter almost—but not quite—completely transparent to neutrinos. There is one exception to this statement that we will discuss later in regard to SN 1987A; another occurs when neutrinos have extraordinarily high energy. This behavior of neutrinos is fundamentally different from the behavior of photons. Because photons interact with matter more strongly than neutrinos do, the more deeply the photons penetrate, the more they will harm living matter. But neutrinos interact so weakly that they can penetrate deeply without causing harm. Indeed, neutrinos don't even "sense" the presence of the matter through which they travel.

This property of neutrinos—the remarkable weakness of their interaction with matter—is of prime importance to the story of SN 1987A, and so another, more realistic illustration of just how weak this property is may be helpful. We stated earlier that a star, say the Sun, generates energy by means of nuclear fusion in its central core. Light—whether particle or wave—travels at 186,000 miles per second $(3 \times 10^8$ meters per second), and the radius of the Sun is 434,000 miles $(7 \times 10^8$ meters). We would expect, then, that if the energy generated in the center of the Sun were to travel as light *in the absence of matter*, it would appear as light at the surface of the Sun in less than 3 seconds. But that expectation is wrong because the photons created in the core of the Sun are scattered and absorbed and reemitted countless times by the matter inside the Sun before they reach the surface. Therefore, a realistic estimate of the time required for that energy to traverse the solar radius from the center of the Sun is on the order of a million years—not a few seconds.

On the other hand, as our earlier thought experiment showed, neutrinos can penetrate even as dense and thick an absorber as the Sun with only a negligible loss of their number. Thus their remarkably weak interaction with matter allows neutrinos created in the Sun's core to travel virtually unscathed directly to its surface in the less than 3 seconds we calculated and thereby to be able to describe the events taking place in the core. That is, neutrinos are messengers that tell us about the events occurring at the center of the Sun.

You may wonder how we know so much about a particle that has little or no mass and no electric charge and that communicates with other matter so feebly. The answer is that it has taken a half-century of difficult experiments at nuclear and particle accelerator laboratories throughout the world to find out all this. What is immediately relevant here, however, is that neutrinos from the Sun have actually been detected on Earth.

Neutrinos from the Sun, or *solar neutrinos* as they are called, were first observed in 1967. This pioneering experiment was initiated and largely carried out by Raymond Davis Jr., then of the Brookhaven National Laboratory, who today still actively continues to collect data from it. The experimental apparatus consists of a sealed steel tank containing 100,000 gallons of a liquid chlorine compound commonly used in dry cleaning. The tank is equipped with stirrers and piping to permit the liquid to be flushed occasionally with helium gas. Once in a great while, a solar neutrino is captured and transmutes a stable chlorine atom into a radioactive argon atom. After about a month, the argon that has been produced is flushed out of the tank by the helium and is collected in a finger-sized tube, which also enables the small number of radioactive argon atoms to be counted. In one month, a nearly unimaginable number (5×10^{19}) of neutrinos—which are capable of converting the chlorine—rain down on the tank and produce roughly 30 argon atoms, of which only about six are actually counted. The extraction and counting operations are repeated regularly each month.

The neutrino detection process is indeed as weak as claimed: only about 30 of every 5×10^{19} (fifty billion billion) neutrinos from the Sun falling on the tank succeed in changing chlorine into argon. Furthermore, the chemical and radiochemical procedures must be very sensitive to allow the collection of a few radioactive atoms from the 100,000 gallons of liquid. To add to the difficulty, the effect of the solar neutrinos is so small compared with the radioactivity that would be produced in the chlorine by the large numbers of cosmic rays that fall unceasingly on the Earth's surface that the tank must be located deep underground. In this experiment, the tank is located in a large room about a mile underground in the Homestake Mine in South Dakota. An artist's view of the mine cavity, the tank, and other equipment is shown in Color Plate 4, which might be thought of as a picture of a neutrino telescope and contrasted with the photographs of the optical and radio telescopes in Color Plates 2 and 3. Strictly speaking, the term *telescope* is usually reserved for an imaging device, which the Homestake detector is not. We stretch the definition here to honor its pioneering achievement.

The observation of neutrino emission from a single star, the Sun, has verified nuclear fusion as the universal energy-generating mechanism in all luminous stars. Without some means of producing energy in the interior of a star, the star would collapse to smaller and smaller volumes under the inward acting force of its own gravity and finally implode. This is one example of nature establishing stable equilibrium through the balance of opposing forces. Consequently, Hans Bethe's identification of nuclear fusion as the source of energy in the Sun was of great importance in the development of a model of the Sun that accounted for its stability. Bethe won the Nobel Prize for this work in 1967, well before the detection of solar neutrinos. Nonetheless, the solar neutrinos exhibit properties completely consistent with their origin in nuclear fusion, and their detection confirmed the nuclear processes occurring in the core of the Sun and also the solar model based on fusion. These conclusions were possible because neutrinos directly sample the events taking place in the core of the Sun and carry essentially undistorted information about those events to us on planet Earth.

Later in this book, we will describe a newer method of detecting solar neutrinos that is radically different from the radiochemical method used in the Homestake detector. It also—because it provides an image of the Sun—is the first neutrino telescope. Future, and more precise, measurements made in this newer way will add to our quantitative understanding of the Sun's interior. For example, solar neutrinos constitute a kind of thermometer embedded in the Sun. Some day, not far in the future, neutrinos will provide a means by which to measure the temperature at the Sun's center—now estimated from theory to be about 15.5 million degrees—and to measure it with a precision of 1 percent.

Another reason for our interest in solar neutrinos is simply the neutrinos themselves. Because the Sun is a copious source of the neutrinos that reach the Earth, we can study their intrinsic properties by indirectly observing their behavior while they traverse the Sun and the long distance to the Earth. Even though we have spoken of neutrinos as particles, they are no more obviously particles than photons are. Both have the same dual wave–particle nature. The energy of each one is specified by Planck's constant multiplied by the frequency of its associated wave. But neutrinos' waves are not made of oscillating electric and magnetic fields, as are photons' waves. The neutrino waves are even more abstract, even less easily defined. Nevertheless, neutrinos and photons are what we call *elementary particles*, similar in their wave–particle natures and their ability to carry energy and momentum, but very different in other basic properties. Although we have a fully developed quantum theory of the photon, we have only an incomplete theory of the neutrino. The reason is largely that neutrino interactions with matter are so uncommon that they require massive, highly sophisticated methods of detection such as the one just described for solar neutrinos. Consequently, we have acquired information about them much more slowly and laboriously than we did for photons, and it is not surprising that our knowledge of neutrinos still is incomplete. In any case, we will study these mysterious particles until their place in the lexicon of elementary particles is clear, and the Sun and other stars are particularly valuable resources for us to do so.

4 LEADING PLAYERS IN NATURE'S BIG AND LITTLE BANGS

It may seem that photons and neutrinos are more similar than not, despite the huge differences in the way they react—electromagnetically and weakly—to matter. And in some important respects they do share certain characteristics. Photons are known to have no mass, and neutrinos have quite small, possibly zero, mass. Both are electrically neutral, and both respond in the same way to the gravitational force, as we shall see later. Finally, both are remarkably abundant products of the "Big Bang," the primal explosion out of which the universe as we now know it probably originated. In this chapter we shall explore the roles of photons and neutrinos in the Big Bang, and their consequent potential influence on the ultimate fate of the universe.

The Big Bang theory of the beginning of the universe is a model of the cosmos tailored to account for three important facts that have been observed in the twentieth century. The first is that the universe—as far out as can be seen with the most powerful telescopes—is expanding. That is, all galaxies (but not Andromeda relative to the Milky Way) have, in addition to their randomly directed motions, a larger component of motion carrying them away from all other galaxies, as if they were traveling outward along the spokes of a wheel. This motion is consistent with the idea of an initial explosion—the Big Bang—from which the elementary particles present at that instant were projected outward at very high velocities. During the period shortly afterward, not more than a few minutes later, these particles combined in ways that led to the formation of the chemical elements with the smallest masses. These newly created elements continued to move with the same high velocities and along the same radial directions as did their constituents. The elements further combined into still larger bodies, thereby slowing down and changing direction, but only somewhat. Consequently, the galaxies farthest from Earth are moving with the highest velocities with respect to Earth, and those closer to Earth have proportionately lower velocities.

A second fact, clearly related to the first, concerns the relative abundances of the natural chemical elements. The Big Bang theory connects these abundances to the earliest formation of the elements, referred to as *nucleosynthesis*. This connection explains the relative amounts of the light elements—hydrogen, helium, and lithium—on Earth, in the Sun, and in other stars.

The third fact is the existence in every cubic centimeter everywhere of a residual very low energy electromagnetic radiation thought to be a direct product of the Big Bang itself. The low energy of this radiation—called *cosmic background radiation* (CBR)—is the result of the universe's continuing expansion since the initial explosion, which has effectively "cooled" the CBR to its currently low temperature, approximately 3 degrees above absolute zero. The initial observation of the CBR and the subsequent meticulously detailed measurements of its properties have lent credibility and respectability to the study of cosmology, in particular to the Big Bang theory. Not so long ago cosmology was thought to be high-class science fiction. But now that we know these three facts and recognize their implications, a new subscience of physics—astrophysics—has come into prominence, serving as a bridge between astronomy and nuclear and particle physics.

As well as confirming the value of the CBR's low temperature, the Big Bang theory also predicts the actual number of CBR photons passing through every unit volume of all space. This number turns out to correspond to the momentary presence (keep in mind that they are just passing through) of 400 photons per cubic centimeter. Furthermore, the theory claims that the primal explosion produced about the same number of neutrinos, which are also to be found today everywhere in all space. Unlike the cosmic electromagnetic radiation, however, the cosmic background neutrinos have not yet been detected because they also have been "cooled" by the expansion and have very low energy. This has further weakened their naturally weak interaction with matter and made them almost impossible to detect. One consequence of this is that the CBR neutrinos literally pass through every cubic centimeter of space, whether occupied by a mountain, a tree, or a human body. Demonstrating the existence of the cosmic neutrino background is a major challenge to modern astrophysicists, one guaranteeing fame to the individuals who solve this fascinating problem.

Apart from the fundamentally different way in which photons and neutrinos respond to the presence of matter, there is yet another major difference between them that has profound implications for cosmology. Photons belong to the class of particles that are not subject to any rules governing their number. In physical processes, therefore, photons are not required to follow any rules other than the conservation of energy and momentum. For example, a photon can disappear within an atom while liberating one of the atomic electrons, or alternatively, an atom excited by collision with another atom may radiate one, two, or more photons. Neutrinos, on the other hand, belong to a class required to satisfy conservation of the number of particles within the class. Accordingly, when a neutrino appears in nature, one of its counterparts must disappear as part of the process, or a companion, called an *antineutrino*, must appear as part of the same production process. In the neutrino world, the neutrino and antineutrino in effect cancel each other and leave their number count unchanged. Conversely, the photon is its own antiparticle and so has no mechanism to conserve its number count. In sum, the total number of neutrinos in nature is controlled, whereas the total number of photons is not. There can be any quantity of photons at any time in any place.

In view of this property of neutrinos, direct observation of the cosmic neutrino background would significantly enhance the credibility of the Big Bang theory. Perhaps more important, that observation would give reality to what may appear to be a fanciful theoretical concept, that the universe is filled with a fixed number of neutrinos streaming in all directions and finding all matter to be transparent to them. The only change that might be envisioned in this immutable sea of neutrinos would be the steady drop of its temperature toward absolute zero as the universe continues to expand.

Is everlasting expansion the fate of the universe? We don't know, but the Big Bang theory suggests another destiny if certain conditions are satisfied. Suppose, for example, that the mass of the neutrino is quite small but not precisely zero. With the existence of so many cosmic background neutrinos, it would take only a very small neutrino mass to make a significant difference in the fate of the universe. Attempts to measure the mass of a neutrino directly in the laboratory have failed because the experiments are not sensitive enough to detect very small mass values. At present, the most sensitive experiments can claim only that the value of the neutrino mass is about 200,000 times smaller than the known value of the mass of the electron which, incidentally, is 2000 times smaller than the mass of the proton. Nevertheless, if we take that value as a rough upper estimate of the mass of a neutrino and if we estimate the volume of space bounded by the most distant stellar bodies observed up to now, then we can calculate the total mass of all the neutrinos in that arbitrarily chosen volume. We need only the additional assumption that 400 neutrinos might be present in every cubic centimeter of that volume at any instant. The result of this calculation is that the total mass of neutrinos is roughly equal to the mass of a million trillion trillion (10^{30}) stars similar to our Sun.

We suppose that a typical galaxy like our Milky Way contains one hundred billion (10^{11}) stars. Assume that they all had the same mass as the Sun. Then our estimate of the total mass of neutrinos would be 10^{20}, or one hundred billion billion, Milky Ways—just because the Big Bang Theory predicts so many of them. Incidentally, this is about all such galaxies that could fit into the volume we have assumed.

These numbers should not be taken literally, as they are guesses of order of magnitude rather than serious estimates. Nevertheless, it is apparent from the result of our overly simplistic calculation that even the minuscule value of the mass of a single neutrino that we assumed would make a significant contribution to the total mass in the universe. Since neutrinos are not visible, they would be a kind of omnipresent "dark matter" in the universe, perhaps weakly bound by gravity to galaxies. Moreover, their presence might help halt the expansion of the universe.

In this alternative to everlasting expansion, all stars (or, strictly speaking, galaxies) would decelerate, then stop, and begin to accelerate inward toward the "origin" of the Big Bang. The universe would subsequently contract into an immensely hot and unstable ball of matter and energy. This unimaginable concentration of matter and energy—far too hot to allow any composites such as atoms or nuclei or even neutrons and protons to exist within it—would in turn repeat the Big Bang cycle. If neutrinos had sufficient mass, as in our illustration, they would play a decisive part in answering the question of whether the universe possesses enough mass and energy to cause it to expand and contract endlessly. Or if it does not have a sufficient quantity of mass, and no other new source of mass or energy is found, perhaps the universe will expand forever, slowly cooling to a frigid hulk.

The thoughts in the last few paragraphs are meant to be provocative but not mysterious. We know so little about the universe that the idea of its either endlessly expanding and contracting or simply expanding for all time should not cause alarm. Furthermore, that the choice between the two alternatives would be made by the amount of mass the universe contained is not a particularly strange idea. Indeed, an analogous situation occurs frequently in everyday life on Earth.

Think of what happens when a ball is thrown vertically upward. The initial upward velocity, which is the result of the throwing action, is slowly decreased by the force of gravity exerted by the Earth on the ball until the ball stops rising and momentarily comes to a complete stop. Then the Earth's gravity accelerates the ball, pulling it downward, and returns it to the thrower. If the throwing and catching were automatic and repetitive, the process might go on endlessly— expansion and contraction. On the other hand, if the ball were thrown with a velocity high enough to enable it to escape the Earth's gravity, proportional to the Earth's mass, it would never return to Earth without further intervention. Our discussion of the fate of the universe thus is no more mysterious than is the example of the ball in the Earth's gravitational field.

5 THE SEARCH FOR EXPIRING PROTONS

Earlier we introduced protons with a minimum of fanfare, although they have always been thought to be the fundamental nuclear particle: smallest in mass and electric charge of all the nuclei, and the immutable and eternal core of hydrogen, the most abundant element in the universe. Nevertheless, it is not unthinkable that protons could contain a deeply rooted residual mechanism for their own spontaneous self-disintegration. This would not be the same mechanism as that responsible for the natural radioactivity of certain of the heavier elements with which we are familiar. The reason is that no fundamental conservation laws are violated when a nucleus decays radioactively; another lighter nucleus and energetic lighter particles are the products of the decay, and energy, momentum, electric charge, and particle type all are conserved in the process. It is not hard to figure out how the first three quantities might be conserved in proton decay, but the particle type in its usual form cannot be conserved in proton decay simply because the proton is the lightest particle of its type and there is no lighter one of the same type for it to decay into.

Why, then, would physicists consider the possibility of proton decay in the face of this prohibition? The conservation laws of physics are empirical, and a new class of phenomena may uncover a previously unsuspected violation. In turn, when this violation, even if minute, is identified and understood, it may be found to have profound implications in many areas of physics. Although the law of conservation of particle type has not been subject to any obvious violations so far, it is not rooted in any deep theoretical precept. Perhaps there is a deeper connection of which we are unaware. One way to find it—or a violation of the law—is to challenge the law with more sensitive experiments. The search for proton decay—if done with sufficient sensitivity—is an experiment that might challenge the law of conservation of particle type with unprecedented sensitivity.

The motivation to question seemingly inviolate principles can be described in another way. In the evolution of modern physics, this questioning has often led to a new unification, that is, to the recognition of deeper relationships that tie together data and theory in previously unimagined ways. These generalizations are a measure of the progress of physics. We next shall look at several of the most recent of them to get a better idea of the vision that lay behind the suggestion that protons might decay, which had, by the 1980s, become enmeshed with the possibility of further unification not only in particle physics but in cosmology as well.

During the early and middle nineteenth century, Michael Faraday and James Clerk Maxwell recognized the seemingly separate phenomena of electricity and magnetism as stemming from a common origin: electrical charges either at rest or in motion. At rest, they give rise to the electrical force; in motion they produce electrical currents and the magnetic force. Currents and magnetism occur in nature both on the microscopic scale of neurons in the brain and in the monumentally charged plasma of the Sun. This unity led to a deeper appreciation of the multivaried phenomena of electricity and magnetism and to the mathematical theory of electromagnetism that we know today. As often happens when unification is especially fruitful, electromagnetic theory became more than a mathematical restatement of the results of experiments: Unification brought recognition that electromagnetic radiation is also due to the motion of electric charges. At the beginning of the twentieth century, this understanding was translated into reality when it became possible to send information in the form of radio waves to a distant receiver. It was initially done, as it still is today, by controlling the motion of electric charges in the antennae of broadcasting systems that transmit radio and television frequencies to receivers everywhere.

Early in the twentieth century, three more great unifying concepts were established. First, the intuitive ideas of time and space were merged into a mathematical world of four dimensions. The idea that time—heretofore always an independent dimension, distinct and separate from space—needed to be treated as a fourth dimension on a par with the three dimensions of space was proposed by Hermann Minkowski and used by Albert Einstein in the first of his two theories of relativity. For reasons having to do with apparent inconsistencies between the 300-year-old Newtonian theory of mechanics and the 40-year-old Maxwell theory of electromagnetism, Einstein was led to devise a theory that resolved these inconsistencies. He did so by unifying time and space and, at the same time, solved the problem of the relativity of motion, an old problem that soon resurfaced in the forefront of the science of the twentieth century.

The theory of the relativity of motion resolves the following dilemma: a physicist measuring an event occurring in nature would get one numerical answer if both she and the event were at rest and a different answer if either she or the event were in motion relative to the other. For everyday low-velocity motions, the numerical difference is so small as to be unnoticeable. But for the high velocities of the then recently discovered elementary particles—electrons and protons—the difference in the answers could be appreciable, even though the same property of the same event was measured in both experiments. For example, an arrow of a given length shot at a very high velocity would appear as it passed an observer standing still as though it were shorter than an identical arrow held in his hand. That would be strange enough, but the real surprise would come from clocks. A clock propelled at a very high velocity past an observer at rest would not read the same time as would an identical clock held in his hand. Einstein's "special theory of relativity"— we shall meet a more general one later—succeeds in accounting for the different results of these experiments and for the seemingly paradoxical results of many other experiments in which the relative velocity of the observer and the event under study is large. In this connection, a large velocity means a velocity comparable to the velocity of light, which in Einstein's theory is the maximum possible velocity of any object in the universe.

The special theory of relativity also provided a second consolidation of seemingly diverse ideas, by unifying the concepts of mass and energy through the famous equation $E = mc^2$, where c stands for the speed or velocity of light. This equation maintains the distinction between mass and energy while making them equivalent and convertible to each other in specific situations. One such situation is the disintegration of a parent radioactive nucleus into daughter particles with high velocities.

We cannot overestimate the importance of the special theory of relativity in modern science. It permeates every aspect of physics, imposing the requirement that certain constraints be satisfied by any result, experimental or theoretical, that claims to be correct. Many in the physics community initially doubted the theory but later accepted it unreservedly. During the

20 years while it was being tested, Einstein was awarded the Nobel Prize (in 1921) for his work on the photoelectric effect, as we noted earlier. The commendation stated that the prize was given without prejudice toward a possible award for his special theory of relativity which was not yet fully confirmed. By the end of the next decade, the theory became so well known and so well accepted, and Einstein's dominance of the subject acknowledged so universally, that his papers on the subject ceased to be referenced in the scientific journals. He had joined the other giants of science. But Einstein never received a Nobel Prize for the special theory of relativity.

At the end of the first quarter of the twentieth century came another development, comparable in magnitude and influence to the special theory of relativity. It was stimulated by a number of experimental results that had been accumulating since the turn of the century but that remained unexplained and undigested. Foremost among these were results that seemed to make sense if light consisted of waves and other, equally respectable results that required light to consist of particles. This is the same paradox that we discussed earlier. Still other experiments using atoms and electrons were faced with the same dilemma: that the two apparently distinct and irreconcilable natures of waves and particles seemed to partake of each other's properties. This dilemma was resolved in an astonishing way by the quantum theory, a product of the outstanding talent in Europe and England before World War II. The word *quantum* here means an amount of a given size fixed by nature. Among the most original contributors to quantum theory were Planck and Einstein, already mentioned, and Niels Bohr, Ernest Rutherford, Louis de Broglie, Erwin Schrödinger, Werner Heisenberg, Max Born, and Paul Dirac, whose combined enormous abilities led to the formulation of a theory with remarkable predictive power that has revolutionized all of modern science. You will find them described in every textbook and every history of modern physics. In the quantum theory, waves and particles are unified, and each does in fact partake of the other's properties in ways that emerge naturally from the theory.

The special theory of relativity and the quantum theory together made modern physics an important influence in intellectual life before World War II. But science did not rest; it expanded its horizons to encompass new discoveries. At the end of World War II, the study of elementary particles and the complex forces among them began and flourished during the following quarter-century. New experimental methods were devised, using very high energy cosmic rays and particle accelerators. Talented young people were attracted to the science's productive past and new challenges. Attention turned to the similarities of the quantum dynamical natures of the electromagnetic and "weak" forces, the same electromagnetic and weak forces mentioned earlier. It was widely recognized that some vital link between them either did not exist or, more likely, had not yet been found. Previous searches for that link had been fruitless, but in 1973 after renewed intensive effort, the missing link was found.

The new experimental information proved that these seemingly distinct forces and the theories describing them were in fact two facets of a single, unified force and of a single theory known now as the *electroweak theory of the combined electromagnetic and "weak" interactions.* In this unified theory, the behavior of both photons and neutrinos in all known electromagnetic and weak phenomena flows from a single source and is quantitatively predictable. Thus, the electroweak theory predicted the exact values of the masses of certain particles involved with the "weak force" before they had been shown to exist. Those values were later found to be correct when, 10 years later, the actual particles were discovered. The generality and elegance of the electroweak theory and its ability to make predictions in accord with experimental results have secured for the theory a leading place in the body of physical knowledge. Furthermore, its success has stimulated a belief in and a quest for a grand unifying theory of all four of the fundamental forces—gravitation, weak, electromagnetic, and nuclear—in order of their increasing strength.

The idea of a grander unified theory beckoned seductively after the success of the unified electroweak theory. By the 1980s, exploration of this possibility was in full swing, and physicists tried to think of phenomena that might, if found, be evidence of a grander unification. These efforts were rewarded by the attractive speculation that perhaps three of the fundamental forces—the electromagnetic, weak, and nuclear forces—may have been equal in strength and character in the early universe soon after the Big Bang and that with the passage of time they grew apart in strength and behavior to become the different forces we now observe. This is an exciting thought, one that launched physics and astronomy in new directions of research and stimulated still greater interest in cosmology. It holds the promise that the study of elementary particles today may lead to cosmological insights into the universe billions of years ago. It would indeed be a grand unification.

These speculations—to the surprise of some and the amused skepticism of others—led to the prediction that protons might not be stable, that they might disintegrate spontaneously into other, lighter particles and thereby constitute a revealing violation of the law of conservation of particle type. This prediction also made allowance for a period between the birth of a proton in the early universe and its disintegration that would be very long, in fact, one hundred billion billion (10^{20}) times longer than the estimated age of the universe.

Again, we are confronted with an enormously large number and must decide what it implies. It is clear that we could not expect to find an average lifetime as long as the age of the universe because then the hydrogen would be gone, or at least going, and so would we. Thus, if protons decay away, they must do so over a period of time much longer than the age of the universe. And if they do so at all, the important issue would be the connection between the elementary particle world of today and the elementary particle world at the beginning of the universe. Furthermore, proton decay—no matter how slow—would signal the nonconservation of a type of elementary particle—the strongly interacting particles—previously thought to be rigorously conserved. It would indicate that conditions in the early universe might have permitted violations of our inviolate physical laws and that we could detect this and other violations if we only looked in the right places. What a triumph it would be to extend our intellectual reach to the earliest history of the universe by means of the elementary particles of today!

Surprisingly, it would not be especially difficult to measure the lifetime of an average proton even if it were so much longer than the estimated age of the universe. This is so because, for example, 10 tons of water contain enough protons (3×10^{30}) so that on average a few of them would decay every year if their average lifetime were approximately equal numerically (say 10^{30} years) to that number of protons. We would need only a large enough tank of water, the proper instruments to detect the less massive elementary particle products of proton decay, and a lot of patience.

The relative simplicity of the prospective experiment and its potential for verifying the overwhelmingly important consequences of the theoretical speculations caused delight and enthusiasm in the physics community. Soon after these speculations were published, scientists tried seriously to carry out the experiment in several specially constructed underground laboratories. In particular, two very large, instrumented detectors were built: one near Cleveland, Ohio, in the United States and one near the small city of Kamioka, in Japan. Each was housed underground in a mine, as was the chlorine solar neutrino detector in South Dakota, to reduce the incidence of cosmic rays. However, the method underlying the search for proton decay was very different from the radiochemical technique used in the solar neutrino apparatus. A schematic drawing of the detector in Japan is shown in Color Plate 5. It holds about 750,000 gallons of water. The tank's 52-foot (16-meter) height corresponds to that of a four-story building. The water purification and recirculation system is necessary to prevent the growth of algae in the stagnant water. Most important, the instrumentation in the water tank, shown in Color Plate 5 as the small circles covering the inner surface of the tank, consists of photosensitive elements and their associated electronics. These elements are able to detect in real time the light emitted during the passage of an electrically charged particle through the water and to determine the point of origin of the particle in the tank as well as its energy of motion and direction.

A photograph of some of the photosensitive elements is shown in Color Plate 6. They are mounted in a supporting structure at the bottom of the tank. The young man in the photograph, Masato Takita, was a graduate student at the University of Tokyo at the time the picture was taken. With all the photosensitive elements in place, as shown in Color Plate 7, the wide-open eyes of photosensitive elements wait to catch and record a ray of light emitted by a particle produced in and traveling through the tank.

The information from each photosensitive element was transmitted electronically to a computer that processed and recorded it on magnetic tape to be checked for format validity and stored for later study. A recorded event consisted of the pattern of photosensitive elements struck by light rays and the strengths of the signals sent from each element to the computer. All the event information was analyzed by a more powerful computer, which determined the particle type and its energy. In addition to the electronic information, it was valuable to have a visual display of individual events that used the abilities of the human eye and brain to grasp complex patterns. The visual display of the information contained in a single event required much human thought and an elaborate computer program.

The design and operation of the detector in the United States was essentially similar to the one in Japan. It had, however, about two and a half times the volume of the detector in Japan and was located in a shallow salt mine off the coast of Lake Erie, as opposed to the location of the smaller detector in a tin-zinc mine in the Japan Alps near Kamioka. It is simpler and clearer to refer in this book to the two detectors by their geographical locations—Kamioka and Lake Erie—rather than the often used acronyms Kamiokande and IMB as in the award certificates at the end of the Preface. The acronyms indicate location and original function: Kamioka N(ucleon: proton or neutron) De(cay) for the Kamioka detector; and IMB for Irvine (University of California at Irvine), Michigan (University of Michigan), and Brookhaven National Laboratory which were the participating institutions for the Lake Erie detector. A schematic representation of the detector under Lake Erie is shown in Color Plate 8. In the further development of the Kamioka detector, several of us at the University of Pennsylvania became members of the Kamioka scientific collaboration. We became familiar with the properties of that detector and participated in the studies made with it, which is why my description of the Kamioka detector and the data obtained with it is more complete than for the Lake Erie detector.

Figure 1 shows the visual display of a proton's possible disintegration as "seen" in the Kamioka detector. Picture this event in the following way: Think of the tank in Color Plate 5 as if it were a can of soup. Use a can opener to free the top of the can, but leave it attached and folded back. Empty the soup from the can, and use the can opener again to free the bottom of the can, and fold it back as well. Finally, split the seam running along the barrel of the can, and fold the barrel back until it, too, is flat. The original three-dimensional can is now represented as a two-dimensional surface, the outline of which is the outline of the display in Figure 1.

The small circles in Figure 1 represent the faces of some of the photosensitive elements in Color Plate 7. The area of the circles is coded to indicate the magnitude of the electronic signal

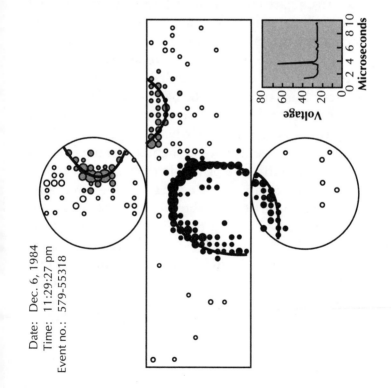

Date: Dec. 6, 1984
Time: 11:29:27 pm
Event no.: 579-55318

FIGURE 1. Visual display (see the text) of an event suggesting the possible decay of a proton. The event has two produced charged particles, indicated by the large circular patterns formed by the small circles representing the individual photosensitive elements in Color Plates 5, 6, and 7. The small circles away from the main patterns are the result of light scattered by the water. The data in the upper-left-hand corner identify the event by number, date, and time. The sharp peak at the lower right indicates that the two particles appeared within a microsecond of each other. When the large circular patterns are analyzed, as shown by the solid line curves through them, they tell the particle type, energy, and direction. This event was a failed candidate for proton decay. Diagram from Katsushi Arisaka, "Experimental Search for Nuclear Decay," Ph.D. diss., University of Tokyo (1984). Arisaka is now a professor at the University of California at Los Angeles.

measured by each element and sent to the computer: the bigger the area is, the larger the signal will be. When elements are struck by the light from charged particles, clear circular patterns are formed, as seen in Figure 1. Each circular pattern has been produced by the light arising from the motion of one of the two charged particles in the tank. It was possible from these patterns for us to find the single point in the tank at which the charged particles originated, to identify the type of each particle, and to determine each one's energy and direction of motion.

From that information we attempted to interpret all the data as a single happening, a single event consistent with the basic laws of physics. If the event was a proton decay, the total energy of all the observed particles had to equal the initial total energy, in this case the mass of a proton at rest. Similarly, the principle of the conservation of momentum constrains the relative directions of the two particles in the event. Events in which the measured values of energy and momentum of the observed particles do not satisfy these conservation laws are judged to have a different origin than proton decay and are set aside. Events showing consistency with the conservation laws, within the measurement errors, would over time provide a small sample of candidates for proton decay to be studied more intensively.

By the beginning of 1984, it was becoming apparent that the number of events qualifying as proton decays was very small, probably zero. This was true of the data obtained from both the Kamioka and Lake Erie detectors. By the end of 1985, it was clear that the theoretical estimate of the average proton's lifetime was too low and that the lifetime—if it were finite—was at least 100 times greater than the predicted value. The euphoria that initially motivated the research was thereby replaced by the hard truth that the secret of the proton's lifetime would not be so easily discovered.

As a storyteller, I might be tempted to end this chapter on the dramatic, seemingly despairing note of the last paragraph. But this would not do justice to the positive accomplishments of the Japanese and American physicists in their search for proton decay at Kamioka and under Lake Erie. All involved were learning how the massive instrumented detectors that they had designed and built actually worked. They coaxed each of the multitude of photosensitive elements in their detectors into steady operation, refined the system to keep the detector water clean and meticulously monitored it, and made sure by frequent testing that the electronic systems were working properly and that the computers were accurately receiving, reading, and transcribing the raw data. These were the necessary conditions just to acquire the data concerning the proton's lifetime.

In addition, these scientists had to write computer programs for data analysis, for visualization of the data to check on the computer, and for simulations of the detector's behavior and also the possible proton decay events as they might appear in the detector. All this was done by small groups of university faculty and students working daily shifts at each detector site while at the same time commuting frequently to their home institutions to teach and visit with their families.

Occasionally in research, as in other walks of life, after a period in which the future appears to be bleak and unpromising, a happy ending appears. This, too, was the fate of the Kamioka and Lake Erie detectors.

6 THE RIGHT METHOD FOR THE WRONG REASON

A subject of occasional contentious discussion in the physical sciences, particularly in elementary particle physics in recent decades, is the degree of influence of theory and theoretical physicists on the choice of goals and plans for experiments. A case in point was the theoretical speculation that protons might be likely to decay over a measurable lifetime, which served as the impetus for constructing the proton decay experiments described in the last chapter. After several more years of collecting data, when the experiments indicated that the average proton's lifetime was a least a thousand times longer than the expected value of 10^{30} years, many scientists decided that these searches—sometimes described as unrewarding, expensive searches for the probably "mythical" proton decay—were a symptom of intellectual weakness in particle physics, a weakness coming from too great an allegiance to abstract theory and too weak an allegiance to hard experimental fact.

There are good reasons that that lower limit on the proton's lifetime, even though it may be as long as 10^{33} years, will not be the end of the story of proton decay. The search will continue despite the feeling that the early experiments were motivated primarily by a theoretical conjecture and despite the acute disappointment when they yielded a null result. But now it is more to the point to see how these underground detectors were diverted from the pursuit of proton decay to another pursuit in which, serendipitously, they played a crucial part in the observation of SN 1987A.

The Kamioka detector was unique in that it possessed especially large area photosensitive elements so that 1000 of them were able to cover 20 percent of the entire internal surface of this four-story high, equally broad detector. This large surface coverage allowed even the small amount of light emitted by moderately low energy electrons to be collected and made feasible the measurement of their energies and directions of motion. It seemed likely that this capability could be extended to still lower energy electrons and perhaps by this means make possible the detection of electrons recoiling from the collisions of solar neutrinos with the atoms in the detector's water. This was a particularly attractive idea because it was known from particle accelerator experiments that a low-energy neutrino would actually collide with one of the atomic electrons and that this electron, recoiling from the collision, would move in the same direction as the neutrino. This might be a method by which the detection of solar neutrinos could be extended beyond the radiochemical technique described earlier, by specifying the source of the neutrinos. That technique, you may remember, exposed the chlorine in the Homestake detector for about a month before flushing out the radioactive argon atoms that were produced, in order to assess their number. In such a procedure, all information relating to the time that a single chlorine atom was converted to argon is lost, and no direct evidence of either the origin or the identity of the particle initiating the transition is available. The weight of the evidence in all the details of the experimental method, as well as the approximate consistency of the experimental result with the expected result based on the theoretical model of the Sun, led to the conclusion that solar neutrinos indeed initiated the events observed in the Homestake detector. Nevertheless, the experiment's inability to point unambiguously to solar neutrinos as the stimuli of the measured radioactivity was a significant weakness in the radiochemical method. Actually it was perhaps more an incompleteness than a weakness, and as something left undone, it was vaguely troubling to Ray Davis Jr., who had begun and carried out the measurements with the Homestake detector. Thus he encouraged and supported the development of any method that might unequivocally resolve this problem.

Unlike the theoretical motivation to search for proton decay, the motivation to observe solar neutrinos in a real-time, directional experiment was empirically based. A real-time, directional experiment implies that the instant during the 24-hour day at which a neutrino interaction in the detector took place must be recorded. This would permit the direction of the observed electron, and consequently its progenitor neutrino, to be correlated with the position of the Sun at that same instant. A positive result in the Kamioka detector would demonstrate that the neutrinos were in fact products of nuclear fusion in the Sun because that was the only process that could account for solar neutrinos of such energy and such abundance. Moreover, tracking the neutrinos back to the Sun would necessarily have to follow the daily rotational motion of the Earth and the Kamioka detector embedded in it. This meant that the neutrinos would arrive at the detector from very different directions as each day wore on. Solar neutrinos would be detected at night in Japan, even though they emerged from the Sun when it was on the opposite side of the Earth. Those neutrinos would have traversed the entire thickness of the Earth to be recorded, in addition to the neutrinos observed during the day when the Sun was more or less directly over the detector.

It seemed almost certain that nuclear fusion was the Sun's energy source and that neutrinos were produced in the fusion reactions. To the best of our knowledge, they had been observed in the detector in the Homestake Mine. It was equally certain that an electron recoiling from colliding with a neutrino would move along an extension of the neutrino's path. Nevertheless, certainty in science is not absolute, and confirmation of a single experimental result, such as the Homestake result, is the proof of the scientific method. When additional tests of what appears to be a well-established result are possible and when such tests might complete and even extend our knowledge, then the opportunity to do so is occasionally too good to miss, especially if it can be done by modifying an apparatus that involved an earlier large investment of time and money.

Accordingly, a collaboration of physicists from Japanese universities, principally the University of Tokyo (whose foresight was responsible for the Kamioka detector in the first place), and the University of Pennsylvania in the United States decided to modify the Kamioka detector with that goal in mind. This required the design and construction of new, sensitive, fast-timing electronics at Penn to exploit another valuable property of the photosensitive elements already in place in the detector. The new electronics improved the measurement of the direction of solar neutrino–induced electrons, thereby increasing the precision with which the neutrinos could be tracked back to their origin. The hardest part of this task was shipping the several racks of electronic equipment from Philadelphia to Kamioka through Japanese customs, which was not prepared for such a reversal of roles between the citizens of the two nations. In addition, the Tokyo physicists took responsibility for large-scale modifications of the detector, chiefly the provision of an electronic and mechanical shield over the entire area of the detector. This shield enabled the identification and measurement of the 30,000 energetic cosmic rays that traversed the detector each day. Without identification and location in time, they were a source of secondary events that could simulate solar neutrino–induced events and give rise to a misleading result.

Preparations for these modifications were begun early in 1984, soon after the initial talks between the senior physicists at Tokyo and Penn. Masatoshi Koshiba and I knew each other only through the scientific literature. He had read my published papers, and I his. We ran into each other at an international meeting in Utah in 1984 and began to talk about various things in science. We commiserated about the negative, disappointing early results of the search for proton decay and went on to speculate about the future of the detectors that had been built for that purpose. There was a meeting of minds and personalities in our talks and the meals we shared. Koshiba had received his Ph.D. at the University of Rochester and had been a member of the faculty at the University of Chicago, and so there was no language barrier between us. We soon focused on the future prospects of the Kamioka detector. He had conceived and built it and was the leader of the Japanese physicists working with it. Our thinking converged rapidly in the next two days, to the extent that when Koshiba, who was scheduled to summarize the results from Kamioka and the plans for its future, found himself with a bad case of laryngitis, he asked me to give the talk for him using his prepared material. I was able to do this easily because my vision of the future so closely paralleled his. Later, we resolved to take seriously the possibility of a collaborative effort at Kamioka and made preparations for visits by Penn physicists to Tokyo and Kamioka.

The visits went smoothly. We discussed at length our common purpose for the Kamioka apparatus during the visits, and our conclusions strengthened our belief that the detector might be able to observe solar neutrinos. Possibly more valuable was the attitude that Koshiba conveyed to his younger colleagues that they should share all their hard-earned Kamioka experience with the Americans. He also underscored the need for complete cooperation between the Japanese and American groups throughout the entire period of collaboration.

The modified detector commenced operation almost two years later, in December 1985. Then the months of learning about and attempting to remove the natural radioactivity in the detector water began. The radioactivity masked the neutrino events that we wanted to observe. This residual radioactivity, usually called *background*, came from very small amounts of uranium and thorium naturally present in the water and, as we ultimately recognized, from the radioactive gaseous element radon. Radon gas readily dissolves in water and produces low-energy electrons as it decays, which closely resemble the electrons that would come from the solar neutrino reactions in the detector water. Radon was present in the air of the mine in sufficient quantity to be a serious problem for the experiment, although not for the health of the experimenters.

From a theoretical model of the Sun and the result of the Homestake Mine experiment, we had calculated the magnitude of the solar neutrino signal that we hoped to observe in the Kamioka detector. We expected that this signal— the number of electrons produced by solar neutrino interactions in the Kamioka detector— would be at most a few hundred electrons in a year. But the observed rate of recorded background events in the detector during most of 1986 was many times larger than that signal. In addition, the background rate fluctuated erratically for no apparent reason, as indicated in Figure 2,

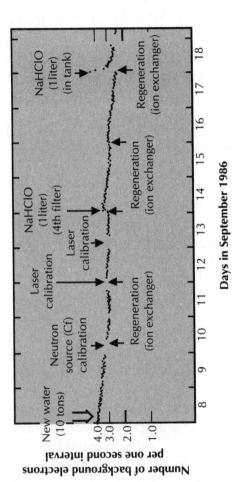

FIGURE 2. The high rate at which the Kamioka detector recorded electrons (entirely background events) over 11 days in September 1986, several months after the modified detector began operation. Each dot indicates the number of electrons recorded in a 1-second interval at a given time on a given day. Notice the several abrupt changes in rate and its slow but significant time dependence. The notations indicate maintenance procedures. Cf is the radioactive source Californium.

which is a small but typical part of the event rate record kept throughout the Kamioka experiment to monitor the performance. In this graphical record, each point represents the measured average number of background events in a 1-second interval at a given time on a given day. The excursions in event rate, seen as the peaks in the plot of Figure 2, occurring at seemingly random times during that period of observation, were a signal that something undesirable that we did not understand was going on in the detector. Until we could come to grips with this problem, we could not make progress toward the goal of the experiment. This impasse continued for many months, and the lack of progress added to the stress on the collaboration of physicists from such different cultures. Patience with the "other group" began to wear thin.

The inherent civility of those involved and their common scientific goal kept the collaboration intact while we tried to solve the high count rate problem. It was very difficult to carry out controlled experiments to learn what was going on in the detector because of its size and unwieldiness. A prime suspect for the fluctuations in Figure 2 was the new electronics, simply because it was new and had traveled so far. This suspicion compounded the tension between the two groups, but many tests failed to produce enough evidence to convict the suspect.

Finally, after a year of frustration, we found the culprit. Each day, about 120 tons of water (5 tons per hour) were taken from the tank, passed through the purification system shown in Color Plate 5 to eliminate radioactive and bacterial matter, and restored to the main tank. As the water came and went, the entire apparatus—main tank and purification system—breathed air in and out (actually, out and in) as if it were a giant bellows. The intake of fresh air from the mine brought with it new radioactive radon, which lives an average of 3.8 days and yields low-energy electrons as one of the products in its decay chain. Once this continuing intake of radon was recognized as being responsible for the observed fluctuations in event rate, it was simple to check it.

When we conclusively demonstrated that radon was in fact the source of the problem, the solution was to seal the apparatus as tightly as possible against the unpurified mine air and to admit only radon-free air during the water recirculation procedure. This was done over several months in a labor-intensive series of improvements. Near the end of 1986, the background event rate from the residual radon in the detector had been reduced by a factor of a thousand, compared with that at the beginning of the year. The event rate then remained stable for several months, as shown in Figure 3, which is a part of the event rate record during a later time period in the course of the experiment. This steady rate, seen day after day in Figure 3, contrasts with the ups and downs in Figure 2. We had at last overcome the principal obstacle between us and our goal.

There were other, though less troubling, events that we needed to resolve. The energetic cosmic rays transmuted oxygen nuclei in the water to radioactive nuclei of boron and nitrogen, which in turn gave rise to electrons that also simulated the electron signal expected from solar neutrinos. Studying these cosmic ray–induced events during the trying times of 1986 taught us to recognize and eliminate them. Then, by the end of 1986, we were able to use them to measure the efficiency with which the detector observed low-energy electrons, thereby turning a problem into a solution.

The ability to cope with these several sources of unwanted, deceitful events, to reduce them to a level of a few events per day, as shown in Figure 3, in almost a million gallons of water, represented a triumph of the experimental method. It was a long step toward vindicating the vision that had

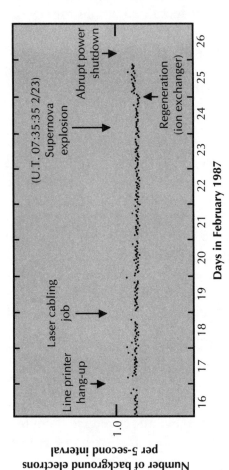

FIGURE 3. The rate corresponding to that in Figure 2 but measured over 11 days in February 1987, during which time SN 1987A was observed. Steps taken before the end of 1986 had stabilized the rate to the low constant value shown, fewer than one count every 10 seconds. This made possible the observation of the supernova explosion on February 23 that is displayed in Color Plate 9. Note also the abrupt, total power shutdown in the mine early on February 26, which is discussed in the text. U.T. stands for Universal, or Greenwich, time.

brought about the conversion of the original detector to its status at the end of 1986. Many physicists were skeptical that the detector, even with its improvements, would be capable of satisfactorily handling what was believed to be the intractable problem of background events. This skepticism, which lurked in the back of all our minds, had to be overcome when we needed the determination to carry on against the background problem. Moreover, these doubts made it difficult for the Americans to find funds to finance their work on the solar neutrino experiment. The scarcity of funds became acute as the yen appreciated significantly with respect to the dollar, at the same time that we were struggling with the background problem. But having solved that problem, our new feeling of confidence made endurable the long wait for the accumulation of a measurable signal of solar neutrino–induced electrons, a signal that we felt sure would show definitively that they had originated in the Sun.

7 THE BURST OF NEUTRINOS FROM SN 1987A

In Park City, Utah, in January 1984, we first discussed modifying the Kamioka detector to make possible a definitive real-time, directional measurement of solar neutrinos. After many months in which the younger physicists paid lengthy visits to Tokyo and Philadelphia and more meetings were held in Tokyo and Kamioka, the actual modifications began. In December 1985, the modified detector started operations, and one year after that, in December 1986, it had settled down to a stable condition with a promising capability for detecting solar neutrinos.

Three years of concentrated hard work, punctuated by arduous travel between Tokyo and

Philadelphia, were necessary to prepare the apparatus and the tools of analysis to carry out the measurements we had initially envisioned in Utah early in 1984. This time scale is more or less the norm for many experiments in modern particle physics and astrophysics. The questions being asked of nature have become more profound, and the experiments correspondingly more difficult than before, requiring a prolonged intensive effort by a larger group of scientists and technical experts. And since the expense is greater, the need for careful detailed planning to avoid failure also is greater. No matter, we were at last ready to begin collecting solar neutrino data.

Then, literally out of the blue, two months later, on February 23, 1987, SN 1987A occurred, as described in Chapter 2. The reaction of the physicists at the Kamioka and Lake Erie detectors to the news of the visual sighting of SN 1987A was compounded by nervous excitement and anticipation and an overwhelming urge to begin reading their magnetic data tapes. Both before and after the visual sighting of SN 1987, every event that the detectors observed was recorded on the data tapes with an identifying number, a precise time, and other pertinent information. Scanning of the tapes began a few days later in both places. In Tokyo we decided to scan the

COLOR PLATE 1 Two photographs of the same portion of the Large Magellanic Cloud, a satellite galaxy of the Milky Way, taken by David Malin and Ray Sharples using the Anglo-Australian Telescope. The photograph on the right was taken in 1984, with the arrow pointing toward the star Sanduleak 202. The photograph on the left was taken in March 1987, approximately one month after the SN 1987A event, which is shown clearly at the position formerly occupied by Sanduleak 202.

COLOR PLATE 2 The United Kingdom Infrared Telescope on Mauna Kea, Hawaii, is a conventional 3.8-meter reflecting telescope, an optical telescope adapted to infrared frequencies. Photograph courtesy of

COLOR PLATE 3 Photograph of the Very Large Array (VLA) of radio telescopes at the National Radio Astronomy Observatory in Socorro, New Mexico. The array contains 27 telescopes, each 25 meters in diameter, located along the three legs of a Y, all of which can be pointed in the same direction. Only nine of the telescopes can be clearly seen here; the rainbow is a bonus. Photograph courtesy of Douglas Johnson, 1981.

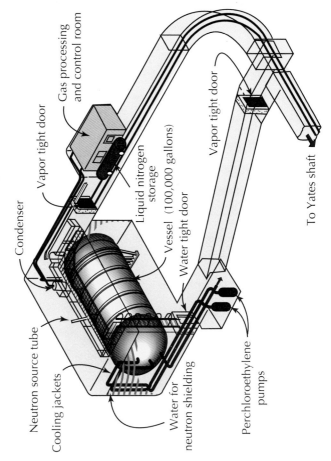

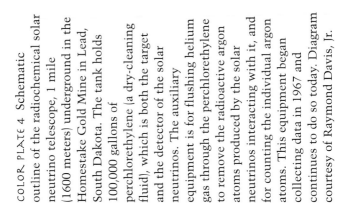

COLOR PLATE 4 Schematic outline of the radiochemical solar neutrino telescope, 1 mile (1600 meters) underground in the Homestake Gold Mine in Lead, South Dakota. The tank holds 100,000 gallons of perchlorethylene (a dry-cleaning fluid), which is both the target and the detector of the solar neutrinos. The auxiliary equipment is for flushing helium gas through the perchlorethylene to remove the radioactive argon atoms produced by the solar neutrinos interacting with it, and for counting the individual argon atoms. This equipment began collecting data in 1967 and continues to do so today. Diagram courtesy of Raymond Davis, Jr.

Gas processing and control room

Vapor tight door

Condenser

Neutron source tube

Cooling jackets

Liquid nitrogen storage

Vessel (100,000 gallons)

Water tight door

Vapor tight door

Water for neutron shielding

Perchloroethylene pumps

To Yates shaft

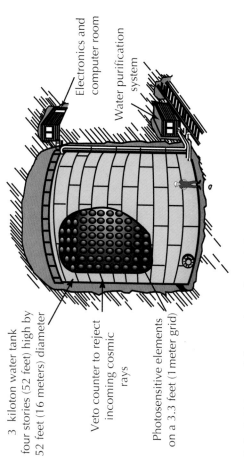

3 kiloton water tank
four stories (52 feet) high by
52 feet (16 meters) diameter

Electronics and
computer room

Water purification
system

Veto counter to reject
incoming cosmic
rays

Photosensitive elements
on a 3.3 feet (1 meter grid)

COLOR PLATE 5 Schematic outline of the original neutrino observatory 3300 feet (1000 meters) underground in the Kamioka Mine in Japan. The tank holds 3000 tons of purified water viewed by 1000 photosensitive elements that detect light emitted by a rapidly moving electrically charged particle in the water. The counting room houses the sensitive electronics that processes signals from the photosensitive elements, and also the computers for preliminary analysis and storage of the data. The water purification system recirculates 125 tons of water each day to remove algae and radioactive matter. The railroad track is for the tram to the mine entrance.

COLOR PLATE 6 Close-up photograph of the photosensitive elements at the bottom of the Kamioka detector, with the water removed. Masato Takita, standing on the support structure, was then a graduate student at the University of Tokyo and is now a staff member at the University of Osaka. The metallic mesh covering the photosensitive elements helps shield against the Earth's magnetic field. The diameter of each glass-enclosed photosensitive

COLOR PLATE 7 Photograph taken with a camera at the approximate center of the Kamioka detector, to show its information-collecting power. The detector is so large that except for the closest ones, the details of the individual photosensitive elements cannot be seen. The high intensity of the light necessary for the photograph produces reflections that are not present when a charged particle is observed.

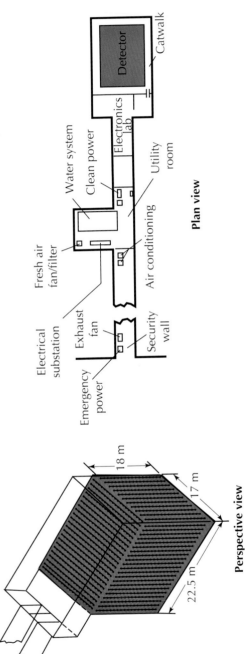

Perspective view

Plan view

Catwalk

Detector

Electronics lab

Utility room

Water system

Clean power

Air conditioning

Fresh air fan/filter

Electrical substation

Emergency power

Exhaust fan

Security wall

18 m

17 m

22.5 m

COLOR PLATE 8 Schematic outline of the neutrino observatory in a salt mine 1970 feet (600 meters) under Lake Erie, near Cleveland, Ohio. Unlike the Kamioka detector, this detector had (it no longer is in operation) a rectangular shape, and its water was contained in a huge rubber membrane that flattened itself against the salt. It contained 8 kilotons of water. The diagram at the right shows the electronics and computer room, the water purification system, and other parts of the laboratory infrastructure. Diagram from D. W. Casper, "Experimental Neutrino Physics and Astrophysics with the IMB-3 Detector,"

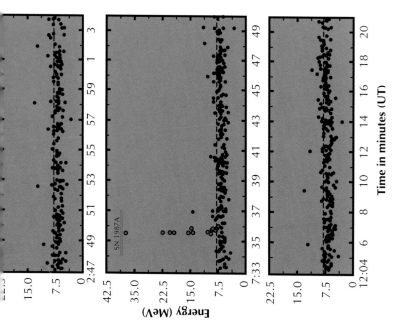

COLOR PLATE 9 Individual events (electrons) observed in the Kamioka detector before, during, and after SN 1987A on February 23. Each dot indicates a single measured event whose energy can be read off the vertical axis (see Appendix 2) and whose time of occurrence is read off the horizontal axis. In the "before" and "after" plots, there is an occasional, single event with more energy than that of almost all the background events, from radioactive elements in the water. The horizontal dashed lines are meant to guide the eye to the region containing most of the background events. The 12 events (positrons) produced by the neutrino burst from SN 1987A in the "during" plot are unmistakable.

COLOR PLATE 10 Photograph of the Hubble Space Telescope in orbit, just after being released from the space shuttle. Courtesy of NASA.
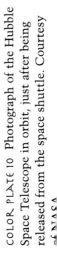

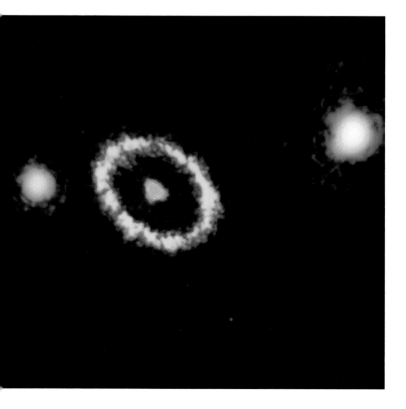

COLOR PLATE 11 A glowing gaseous ring seen around the tightly packed remnant of SN 1987A in August 1990, which appears as the red blob near the center of the ring. The blue stars are not associated with SN 1987A. The photograph was taken with the European Space Agency Faint Object Camera and the Hubble Space Telescope. Courtesy of NASA.

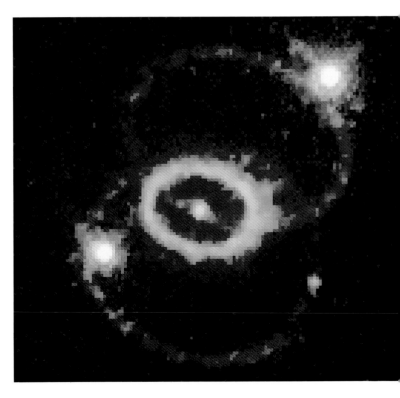

COLOR PLATE 12 Structure of three glowing gas rings seen around the remnant of SN 1987A in February 1994. The yellow orange ring and the blob near its center are the same as in Color Plate 11, but the two larger rings were not present earlier. The yellow orange ring is in the plane containing the supernova, with the two larger rings in front and behind that plane. The photograph was taken with the Wide Field Planetary Camera 2 and the Hubble Space Telescope. Courtesy of NASA and Dr. Christopher tapes by beginning a number of hours before the earliest reported visual sighting and proceeding in 10-second intervals, always starting the next interval 9 seconds after the start of the previous interval, to ensure a complete overlap in the course of the scan. Everyone was afraid that even if a signal from SN 1987A had come in the form of neutrinos—which was not at all certain—the signal might be missed because it was too weak or too short or too long in duration. In this connection it is interesting to reflect on a tacit assumption in the experimenters' thinking. All of us knew from the visual sighting of SN 1987A that it lay in the Southern Hemisphere and also that our detectors were in the Northern Hemisphere. For neutrinos from SN 1987A to reach the detectors, the neutrinos had to penetrate a large fraction of the Earth's diameter. No one doubted their capability to do so. If we did not observe any neutrinos, our thinking went, the fault would be due to misconceptions relating to supernovae, but not to our ideas concerning the weak interaction of neutrinos.

Of course, things are rarely as clear-cut as that. There is also a well-known, different type of supernova explosion from which neutrinos would not be expected and that might or might not be recognized by other properties as being of that different type. A failure to observe neutrinos from SN 1987A could be traced to basic misconceptions about supernovae or simply that SN 1987A was of the wrong type to do so; only the actual detection of neutrinos would be definitive.

To make sure that every event was scrutinized carefully enough, the identification number, energy, and time of occurrence of each event were printed out automatically as the computer read them from the tape. After a few hours, the overworked printer overheated, and the scan was stopped while a substitute printer was found and installed. Our excitement was palpable. Several hours later, the burst of neutrino events was found. Contrary to our fears, it was a clear signal, standing out "like a sore thumb," with no doubt as to its origin.

The earliest recorded signal from SN 1987A was not the light seen on McNaught's photographic plates, but a burst of 12 neutrino-induced events in the Kamioka detector and eight similar events in the Lake Erie detector. All 20 events were observed within a few seconds of one another at 0735 hours Greenwich Time on February 23, roughly 3 hours before the first optical sighting. The 12 events in the Kamioka detector came in the short time period of 12.4 seconds.

The eight events in the Lake Erie detector occurred during the same time period and had similar energies. All the events in both detectors were low-energy electrons—actually, as recognized later, low-energy positively charged electrons, the anti-electrons called *positrons*. Coincidentally, those in the Kamioka detector had energies in more or less the same range as those expected for solar neutrino–induced electrons; those in the Lake Erie detector were somewhat higher. Without the 3 years spent modifying the Kamioka detector and a similar, shorter period in which the sensitivity of the Lake Erie detector was improved, the low-energy positrons produced by the SN 1987A neutrinos might well have never been noticed.

We now know that this burst of neutrinos came during the first few seconds when a neutron star was formed within SN 1987A and that the light seen 3 hours later came from the excited outer mantle of the exploded progenitor star, Sanduleak 202. You may wonder, then, why we started the story of SN 1987A with the visual sighting. The answer to that question comes from our earlier description of our method of acquiring and storing data in the underground water detectors. Event data, including an event identification number and the date and time of its occurrence, were recorded on magnetic tape in the counting room near the detector in the mine (see, for example, Color Plate 5). Each tape held the information from many events, which filled two tapes every day. The tapes were not read or analyzed in the mine. Instead, each week a student loaded them in a backpack and walked the quarter-mile underground to the mine station where he boarded a train for the trip to the mine entrance. Outside the mine, a short car ride brought him to the common living quarters. These were small apartments for some of the miners and their families with a communal dining hall and communal bathing and laundry facilities. There, the Tokyo and Penn groups each had an apartment where they packed up the tapes for express delivery to Tokyo and Philadelphia.

There was no compelling reason to speed up this procedure while we were searching for proton decay or solar neutrino events. Only the accumulation of data over many months would be useful in those searches, and nothing of special importance was expected on any particular day. We had discussed the unlikely possibility of a neutrino burst from a supernova, but the extra effort and expense required to respond promptly to this unlikely possibility did not compete at the time with the more pressing need to understand the detector's behavior.

I am unable to resist inserting a personal note here. While all this hard work was being done in Tokyo, I was in Philadelphia taking care of my teaching duties at Penn. I was aware of the visual sighting of what was presumed to be a supernova and also keenly aware of the struggle in Tokyo to scan the data tapes. Tense and irritable with anticipation, I did my best to remain calm. However, as usual, several members of the Penn group were in Tokyo. Within an hour after the clear signal of the neutrino burst was obtained, one of the Penn graduate students, Soo Bong Kim, kindly telephoned me (e-mail was not available 10 years ago) to describe in detail the observation. He quoted the times at which the positrons were observed, their energies and directions, and the event rates in the detector before, during, and after the neutrino burst. Although my first instinct was skepticism prompted by scientific conservatism, I could find no obvious fault with the "sore-thumb-like" signal.

My immediate impulse was to get on the next plane to Tokyo. We needed more study of the events in the burst; I wanted to talk with our collaborators about the implications of the burst, and we had to write a technical paper describing the observation and send it off to a journal. As I prepared to make a plane reservation, I suddenly remembered that I had promised to attend the opening night of a play in Philadelphia in which my daughter, a professional actress, was starring. If I were to go, I would have to delay the trip to Tokyo for two nights. What should I do?

Now, many years later, I am glad to be able to say that I kept my promise and stayed to see the play. I satisfied my desire to be involved with the data in Tokyo by means of the telephone. This provided the facts, though not the substance, of the discovery. While still in Philadelphia, I had little opportunity to savor the sweet taste of the discovery and to celebrate with my collaborators. A self-imposed silence added to my discomfort, as I was unwilling to speak to my colleagues at Penn, even confidentially, before the data were probed for possible errors and a description of the discovery was written. And all the while I wondered—unwilling to ask—which, if any, of the other neutrino detectors in the world had made a similar observation.

There is a small moral to be drawn from this anecdote: let no one tell you that science is a dispassionate, emotionless pursuit. It is as emotional and demanding of the psyche as any other creative human endeavor.

The observed neutrino burst was the first ever neutrino signal from a supernova. Before our nervous energy could drain away, we studied the events in the burst more carefully still, tabulated and plotted their properties, and prepared a brief report to the *Physical Review Letters*. By then I had gone to Tokyo. Several astrophysicists in the United States, Japan, and elsewhere had theorized earlier that neutrinos would be emitted from a supernova of the type that SN 1987A turned out to be. So we thought it appropriate to telephone each of them and read to them the abstract of the report we had just prepared so that they would be the first to hear of our observation. We did this and also fulfilled their requests for more and more information. At the same time, we notified the physicists at the Lake Erie detector of the number of events detected and the times of their occurrences. The Lake Erie detector was less sensitive than the Kamioka detector, because its area covered by photosensitive elements was about half as large as the corresponding area in the Kamioka detector. Consequently, the Lake Erie physicists needed to know the exact time interval in which they should look for the burst. They searched that short interval and, to everyone's relief, found the signal of eight events. The observation of a neutrino burst from a supernova was soon common knowledge throughout the world's scientific community. The telephone in the Tokyo laboratory rang constantly with requests for confirmation of the rumors that were circulating, and Japanese and foreign reporters besieged us with requests for interviews.

Two smaller, less sensitive detectors, one in the Mont Blanc tunnel between Italy and France and one under a mountain at Baksan in Russia, also were operating on February 23, 1987. The Mont Blanc detector reported a weak neutrino signal several hours earlier than the Kamioka and Lake Erie signals, which raised the possibility that more than one neutrino burst had come from SN 1987A. This roused some scientists and media people to a frenzy. But the physicists at the other detectors diligently searched their data for the time specified by the Mont Blanc data and could not find an earlier signal (or a later one). Ultimately it was decided that the early signal was probably a larger than usual statistical fluctuation that momentarily misled the Mont Blanc physicists, just as a larger than usual fluctuation of the stock market occasionally misleads investors. This episode had a double-edged irony because the Mont Blanc detector had originally been intended to search for neutrinos from a stellar collapse. What a disappointment it must have been for the scientists who had planned and constructed that detector to have missed the neutrinos from SN 1987A when other detectors observed them more or less by chance. Sometime later, the Baksan detector reported a small signal of four events coinciding in time with the Kamioka and Lake Erie signals. Thus the furor stimulated by the possibility of more neutrinos than we could explain died away peacefully. What could have been contentious competition was resolved, and a way was opened to new research in many areas of physics and astrophysics.

The brief private period of discovery came to an end with the technical report safely in the mail to the *Physical Review Letters*. The exhausted Kamioka physicists celebrated the occasion with an elaborate Chinese meal, followed by a good night's sleep.

It is not absolutely necessary that you be familiar with the details of the observation that constituted the neutrino burst, but you may

understand better the nature of astrophysical research in general and the burst observation in
particular if you look at the actual data. The rates
of recorded events observed by the Kamioka detector before, during, and after the neutrino burst
are shown in the three plots of Color Plate 9. In
these plots, the vertical axis measures the event
energy, and the horizontal axis measures the time
in minutes. Unlike the points in Figures 2 and 3,
each point in the plots of Color Plate 9 represents a single observed electron or positron. Its
energy is read off the vertical axis, and the time
at which it was detected is read off the horizontal axis. (The units in which energy is measured
are briefly discussed in Appendix 2.) The concentration of electron events below the dashed
line in the plots in Figure 3 is due to the residual
radioactivity in the detector water; it was never
fully eliminated. The "before" and "after" time
intervals in the top and bottom plots in Color
Plate 9 show the detector's quiescent state—the
steady-state number of events from radioactive
sources—as it was at any hour on any other day
in late 1986 and 1987. These plots contain the
raw event-by-event data from which a plot over
a much longer time interval, such as that in Figure 3, is constructed. In the "during" interval, the
burst of 12 positron events within 12.4 seconds

stands out as an unmistakable, unique happening. It is no wonder that all of us became almost instant "believers" when those events, with such relatively high energies and in such a relatively short time interval, emerged from the computer and were printed for us to read.

Much later, we were able to combine the burst data from the Kamioka and Lake Erie detectors in a single plot. This is shown in Figure 4, in which each of the 20 data points is interpreted, as are the points in Color Plate 9, as a single positron detected at the indicated time with the indicated energy. The vertical bar attached to each point indicates the uncertainty in the measurement of the energy of that event. Combining the data from the two detectors entailed a small difficulty that proved to be mildly embarrassing to the Kamioka physicists. The site of the Kamioka detector had no absolute time-measuring device, such as an atomic clock, that, when calibrated, would keep accurate universal (Greenwich) time over a long time interval. We did have a quartz-based oscillator clock in the computer, but this tended to drift by a few seconds owing to temperature changes in the electronics hut and consequently required frequent checking (calibration) against a more stable, absolute clock such as one available at most of the

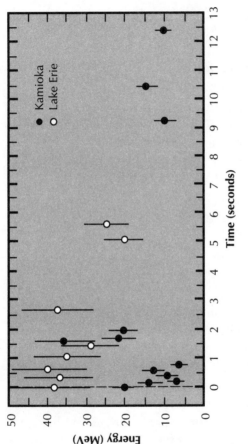

FIGURE 4. Plot showing the 12.4-second duration of the neutrino burst and the energies of the produced events from both the Kamioka and Lake Erie detectors. This plot is similar to the "during" plot in Color Plate 9, but with a time scale in seconds rather than minutes. The earliest measured event in the Kamioka sample is adjusted to coincide in time with the earliest event in the Lake Erie sample, for a reason discussed in the text. The vertical bars through the data points indicate the approximate errors in the energy measurements.

national standards laboratories in the world. As long as the electrical power remained on in the electronics hut, it was possible to check the computer clock against the standard clock as often as we desired, either before or after the occurrence of an interesting event. To our horror, however, there was an abrupt and total power shutdown in the mine early on February 26, which erased all memory of past times in the computer (see Figure 3 and the notation for that date).

Each day, the physicist on shift at the Kamioka detector would telephone the Japan standards laboratory for the correct time, accurate to better than a second, and enter that time into the computer. The entire procedure took at most a few seconds, since it had become habitual for all concerned. When the technical paper was being written in Tokyo without a post–SN 1987A calibration of the computer, we looked to see who had been on shift at the time of SN 1987A and asked him to estimate the uncertainty in the time he had recorded. More or less casually he reckoned "much less than a minute," and equally casually we stated a time uncertainty of 1 minute in the paper that went to the journal and was published remarkably soon thereafter. We had been so casual in our approach to the precise time because we had the mistaken impression that the Lake Erie detector had been temporarily shut down and that we had no precise time correlation with another detector. But it turned out, fortunately, that we were wrong about the status of the Lake Erie detector, which had been mostly, but not completely, turned back on. Also, in the intense excitement of the moment, we had not paid sufficient attention to the Mont Blanc and Baksan detectors or to the solar neutrino detector in the Homestake Mine. The Homestake detector had not observed a neutrino signal from SN 1987A at the time of the Kamioka and Lake Erie observations, nor with hindsight could it have been expected to have done so, considering its small size.

The physicists at the Lake Erie detector properly quoted a few seconds' uncertainty in their estimate of the time of SN 1987A while at the same time, we at Kamioka, with red faces, tried to explain the origin of our estimated, much larger, uncertainty. The times actually recorded for SN 1987A at the two detectors agreed within a few seconds; hence the combined data in Figure 4 are plotted so that the first event in the Kamioka data is adjusted to coincide in time with the first event in the Lake Erie data.

It may appear that the competition between the Kamioka and the Lake Erie detectors was a distraction and perhaps even a source of contention exacerbated by the presence of the smaller detectors. Once, however, the emotional high of the discovery dissipated, we all recognized that the confirmation provided by the two experiments half a world apart made for instant acceptance by the scientific community of the validity of the neutrino burst from SN 1987A. Indeed, one might wonder about the reaction of astronomers and physicists if events similar to those in Figure 4 were to be found in the future during the normal reading of data tapes and in the absence of a visual sighting of an exploded star. Would such a burst of neutrino events be accepted as a signal from a future supernova? This is a question of much interest because a future supernova in the Milky Way, in which we expect only a few supernovae each century, may be located so that its direct light would be blocked from reaching Earth. But as we have seen, such a blockage would not prevent the neutrinos from reaching us. The neutrino burst accompanied by the visual sighting of SN 1987A made a compelling case for the validity of both the neutrino and visual observations and their interpretation.

SN 1987A will lend credibility to a future observation of a presumed neutrino burst without a visible pointer to a progenitor star. But for the burst's acceptance as a fact, confirmation by two or more detectors, each with clear signals at precisely the same time, will be necessary, as it was in the case of SN 1987A.

8 WHY ALL THE EXCITEMENT?

The Kamioka and Lake Erie physicists had hoped after a few days of rest to consider the many implications of the neutrino burst. The two groups' initial publications were limited to describing the observations of the burst and the properties of the individual events. Once we had sorted out our ideas, we had planned to discuss the implications in papers soon to follow the first. In Tokyo, for example, we had begun out-lining subsequent papers and made tentative as-signments to those individuals responsible for the first drafts. But this was not to be. When not sleeping or eating during the long trip from Tokyo back to Philadelphia, I worked diligently on my assignment, only to find upon reaching my office, copies of several manuscripts on the subject of my assignment already submitted for publication by others outside the Kamioka and Lake Erie collaborations. In the remainder of that week and the following week, copies of more manuscripts arrived, anticipating all our plans for publication. The explosion of Sanduleak 202 and the detection of neutrinos marking the earliest moments of the explosion gave rise to an instantaneous flood of technical papers analyzing and reanalyzing both the optical and the neutrino data. The implications for astronomy and physics of the neutrino burst were explored in literally hundreds of talks and articles. SN 1987A even made the cover of *Time* magazine because it explained so clearly, even to nonscientists, the evolution and ultimate death of a massive star and the simultaneous birth of a neutron star. On the other hand, to elementary particle physicists the information about the neutrino's properties was paramount. There was hardly an opportunity for the Kamioka and Lake Erie experimenters to get a word in edgewise.

The reason for all this excitement was that the neutrino burst brought together previous ideas—based on sound theory, hard fact, and considerable speculation—into a single, largely complete and correct description of the supernova process and its cosmic implications. At the same time, the emission of neutrinos from the enormously dense core of Sanduleak 202 and their voyage to Earth from that star in the Large Magellanic Cloud exposed certain properties of neutrinos with a clarity never before available. The full spectrum of meaning of SN 1987A for the different scientific disciplines, as well as for nonscientists, made it an object of worldwide interest from the very moment it happened.

Several of the most remarkable features of SN 1987A transcend the many detailed technical interests of the scientists. These features can be described simply, without recourse to elaborate mathematical formulation, because they are basic to the general nature of the supernova phenomenon. Likewise, in discussing them we can convey a sense of the magnitude of the supernova phenomenon, to show how it is governed by the laws of physics and astronomy, and to point to the ways in which it is consistent with our understanding of other natural phenomena.

For example, we made the plausible assumption that SN 1987A emitted neutrinos equally in all directions, so that the detectors intercepted only the tiniest fraction of them. In addition, we assumed that the SN 1987A neutrinos interacted with matter—the water of the Kamioka and Lake Erie detectors—as do the neutrinos produced in the accelerator laboratories that we physicists have studied for many years. Then, using the data for the 20 events in Figure 4 and the known distance from SN 1987A to Earth—about 160,000 light-years, or approximately 10^{21} meters—we can directly estimate that in total, SN 1987A emitted 10^{58} (ten to the fifty-eight) neutrinos. How many are 10^{58} neutrinos? Written in words, it is ten thousand billion billion billion billion billion billion neutrinos. To put this in perspective, if you consider that there are about 5 billion people on Earth now, that people are heavily outnumbered by insects, and that insects in turn are heavily outnumbered by bacteria and viruses, you may begin to get a sense of the meaning of all those billions that make up 10^{58} neutrinos. Incidentally, 10^{16}, or ten million billion, of them fell on the Kamioka detector, of which, as we said, 12 of them were actually detected. What better confirmation of the weakness with which neutrinos interact with matter!

In any context, even in astrophysics, 10^{58} is a large number. It is roughly the number of neutrinos that the Sun will have produced during its entire lifetime. It is a remarkable number when it denotes neutrinos with the average energy indicated by the events in Figure 4. We can estimate, again directly, that SN 1987A generated far more power during its 12 seconds of emitting neutrinos than have even the most powerful stellar objects—the Active Galactic Nuclei—in a similar time interval. Furthermore, with regard to power generation by humans, during that 12 seconds, SN 1987A generated an amount of power in the form of neutrinos that dwarfed all power production on Earth by about 34 powers of 10. You can see now why we think of a supernova as a cataclysmic event.

A final note on the energy carried by the neutrinos: It exceeds by at least 100 times all the other energy emitted by SN 1987A, the energy in the visible light seen in Color Plate 1 coming from the excited mantle of Sanduleak 202 and the energy in all the incredible sound and shock waves generated by the collapsing core. How ironic that nature turns to the two fundamental forces of least intrinsic strength—gravitation and the weak force—to create a phenomenon of such enormous energy.

Moreover, if you recall the thinking in Chapter 4 and speculate on the number of supernovae that have occurred in the roughly 15 billion years since the Big Bang, you will realize that—even apart from the neutrinos emerging from the Big Bang—the neutrinos produced by all past supernovae also contribute to the neutrino content of the entire universe and perhaps to its mass and energy content as well.

Nevertheless, we need not be intimidated by these extremely large numbers. SN 1987A was a natural phenomenon, and it can be understood as such, with our knowledge of how stars are formed, evolve, and decline. In doing this we need also to introduce certain basic aspects of nuclear physics that dominate the supernova process. This will reveal how such a phenomenon could have generated so vast an amount of power in the destruction of Sanduleak 202 and its aftermath.

9 BIRTH, EVOLUTION, AND DECLINE

Ideas about the birth of a star are plausible but not completely conclusive. Generally speaking, we think of a concentration of hydrogen gas rising randomly in the cosmos and slowly transforming—by the mutual gravitational attraction of one molecule to another—from a loosely connected assembly to a more compact mass. With time, the diameter of the compacted object—not yet a star—grows smaller and smaller and its temperature steadily increases as gravity shrinks the volume occupied by the hydrogen. At this stage the hydrogen does not generate an outwardly directed force to balance the inward force of gravity, and the hydrogen mass is not large enough to precipitate a runaway collapse under the action of gravity. The would-be star's size decreases and its temperature increases until, as we pointed out in Chapter 3, the temperature reaches a value at which nuclear fusion can begin. The energy developed in nuclear fusion—with the release of energy, four hydrogen atoms are joined as helium—provides the outward force necessary to stabilize the infant star against the implosive force of gravity. Some of that energy ultimately reaches the surface of the star, causing it to shine. These loose ideas of star formation describe the stars born today as well as those born at other times in the history of the universe.

Although our description of star formation is brief and primitive, it is consistent with what we have known for three centuries about gravity and what we have learned about nuclear fusion in the last 35 years. For example, you may ask how we know that a certain critical or threshold temperature is required for nuclear fusion and how we know the magnitude of that temperature. The answers to these questions come from laboratory experiments. At a relatively low temperature, the electron in a hydrogen atom can be freed from its orbit around the proton. For nuclear fusion to begin, a pair of protons must get close enough to each other to be able to fuse. Each proton carries a positive electric charge that is neutralized by the negative electric charge of the electron when it is present but that is free to act when the electron is removed. At low temperatures, the mutual repulsion of their like charges keeps the two protons far enough apart to prevent them from fusing. As the temperature of the unborn star rises, the protons move with increasing speed that occasionally enables them to penetrate the electrical barrier that keeps them apart. Incidentally, this effect is forbidden in classical theory but is allowed in the more general quantum theory. We stated earlier that the temperature at the center of the Sun, where we know that fusion is taking place, is calculated to be 15.5 million degrees. This matches what we know from laboratory experiments, in which we proved that extreme heat is necessary for fusion to occur in a star. Note that when we refer to a temperature value in millions of degrees, the temperature scale is technically the "absolute" scale, with its zero value at −273 degrees Celsius.

Most of the detailed questions about fusion can be answered from information gained in laboratory experiments. Questions on a larger scale or of a more profound nature, however, have no easy answers. We may ask whether the populations of the stars, galaxies, and clusters of galaxies are in fact randomly located in the cosmos, since we assume that initial random concentrations of hydrogen give rise to the stars. There is good evidence that the larger objects—the galaxies and clusters of galaxies—are not randomly distributed, that they may exhibit a regularity of structure on a cosmic scale. But we cannot yet confirm this definitively.

A more difficult question concerns the origin of the randomly located concentrations of hydrogen that began the formation of stars. We can ask essentially the same question about other seemingly random origins, for instance, the origin of life. It is thought that randomly located concentrations of amino acids on Earth in the prebiotic era enabled protein synthesis to take place and primitive life forms to emerge. In discussing both origins, of stars and of life, the initial premise is an arbitrary one, the arbitrariness conveyed by the word *random*. It is a word that provides a seemingly objective, statistical cloak of scientific respectability for the discussion of origins, and it hides our ignorance—perhaps temporary, perhaps permanent—of the precise starting conditions.

A star evolves with the passage of time, as do all created things, but evolution in astronomy is very different from evolution in biology. In astronomy, individuals change species as they evolve or age, but species remain the same. In biology, on the other hand, individuals remain within one species as they age, but species change or evolve with time. In the case of a star, its lifetime may be as short as a few million years or as long as several billion years, during which the continuing process of fusion depletes the star's initial store of hydrogen. At the same time, hydrogen may be accumulated by means of the gravitational attraction of ambient matter outside the star. The result may be either a net increase or a net decrease of the star's mass and fuel supply, depending on its location and its environment. It is tempting at this point to go beyond this recital of scientific facts to compare the maturing star with a maturing human or other living being, continually expending its internally generated energy to survive and being buffeted by the winds of change as it ages. This may not be altogether a foolish, sentimental thought. Science assumes the universality of nature, the idea that seemingly disparate things and phenomena can be understood in terms of the unifying principles governing their behavior. We divide our world into the living and the nonliving, based on the pragmatic definition that to be alive is to be born, to ingest fuel, to generate energy from it, to eliminate the resulting waste, and, finally, to die. Can we apply this definition to the life of a star, especially when we shall soon see that a star does indeed die? We leave this thought as an incentive to question our perhaps too limited view of the universe.

Sooner or later, a star uses up the hydrogen in its core, and then it turns to its other resources in an effort to survive. As its vitality wanes, the star once again is compressed into a smaller, hotter sphere by the force of gravity, just as it was in the early stage of its birth. But the star does not yet implode under the action of this compressive force. Rather, its temperature rises significantly until its internal furnace becomes hot enough to begin a new round of fusion, in which the product of the earlier fusion—helium—is converted to carbon and oxygen. This round requires a higher temperature because helium nuclei—with two protons and two neutrons—are doubly charged electrically and a higher temperature is necessary to bring the helium nuclei into close enough contact to allow them to fuse. With the burning of this new nuclear fuel, the core of the star again generates energy, and equilibrium is once more established between the compressive force of gravity and the expansive force of the new round of fusion. This does not last long, however. In several tens of thousands of years, the helium itself is exhausted, and the contraction and heating begin again, this time leading to the fusion of carbon and oxygen—the products of the last round—to produce the element silicon. In a few thousand years, the carbon and oxygen, the fuel of the third round of fusion, also are used up.

You may assume that the star will continue to burn all the available fuel until—like wood burning in a fireplace—nothing is left but ash. You would not be far wrong, providing that you recognize the fundamental difference between the low-temperature, chemical burning of a wood log and the extremely high temperature burning of nuclear fuel. That is, the burning familiar to us in everyday life is a chemical phenomenon, whereas the burning of elements in the furnace of a star—in which heavier elements are produced from lighter ones by means of fusion—is a nuclear phenomenon governed by constraints imposed by the laws of nuclear physics.

First, the constituents of all atomic nuclei are electrically neutral neutrons and positively charged protons, approximately equal in mass. Second, the nucleus of the common form of hydrogen is solely a proton, but nature also allows a heavier hydrogen nucleus, the deuteron, which is made up of a neutron and a proton and is stable against spontaneous disintegration. The neutron and proton in the deuteron are held together by the strongest of the fundamental natural forces, the attractive nuclear force between them. Gravitation is too weak to hold them together, and the neutron's electrical neutrality eliminates electromagnetism. The Fermi, or "weak," force discussed earlier is also incapable of holding together a neutron and a proton.

The third constraint arises from the fact that the nuclear force acts only when the neutron and proton are relatively near each other, unlike gravitation and electromagnetism which can act over longer distances. This means that a constituent neutron or proton may be completely freed from a nucleus if it is given sufficient energy to escape the short range of the nuclear force. Fourth, as we have already seen, the electromagnetic force causes two protons to repel each other and prevents a stable nucleus of two protons.

Fifth and last, in a stable nucleus the nuclear force acts with equal strength between all pairs of constituents—protons with protons, neutrons with neutrons, and protons with neutrons. In nuclei with more than a single constituent, the energy with which each neutron or proton is bound to the others—called the *binding energy per constituent*—is the total energy of all the pairs holding the nucleus together divided by the number of constituents. This accounts for the much tighter binding of the four constituents—two neutrons and two protons—in the helium nucleus, compared with the binding of the two constituents in the deuteron. In helium, the binding energy per constituent reaches a value almost equal to the value in heavier nuclei with more interacting pairs but also with more constituents. To liberate a constituent from the nucleus requires an amount of energy equal to the binding energy per constituent. The liberating energy may be supplied in the laboratory by the collision of the nucleus with a high-velocity projectile—which may be some elementary particle or other nucleus—produced, say, in a particle accelerator. Or on a cosmic scale, the liberating energy may be supplied by heating the nucleus to a high enough temperature to free a constituent; recall the temperature in the center of the Sun, for example.

When we look at the elements with increasing numbers of constituents, going from helium with four constituents to iron with 56, we find that the binding energy per constituent also increases, but much more slowly than the increase in the number of constituents. Which of the constraints accounts for this relatively slow increase of the binding energy per constituent? The answer lies in the fourth constraint. Nature tries to keep the numbers of protons and neutrons in a nucleus equal, or at least approximately equal. For example, the deuteron has one proton and one neutron; helium has two of each; carbon has six protons and six neutrons; oxygen has eight of each; and so on. But in accord with the fourth constraint, the protons in a nucleus act through the electrical force to repel one another. When only a few protons are present, as in the lighter nuclei with small numbers of constituents, the total binding energy is dominated by the attractive nuclear force, and the repulsive electrical force is subdued. The total binding energy is reduced by only a relatively small amount because of proton repulsion in the elements lighter than oxygen. When, however, the number of protons is larger, as in the heavier nuclei, electrical repulsion of the protons substantially reduces the total binding energy. To compensate for this effect, the heavier nuclei contain more neutrons than protons, and therefore more constituent pairs, so as to take advantage of the fifth constraint. Thus, iron has 26 protons and 30 neutrons.

Despite this attempt at compensation, the binding energy per constituent slowly decreases from its value in iron, the highest value of all nuclei, until it reaches its lowest value, which is in uranium, the heaviest stable nucleus with 92 protons and 146 neutrons. You may wonder why nature does not simply add more neutrons to the heaviest nucleus, which might then overcome the electrical repulsion of the protons and enable still heavier nuclei to be formed. But natural order, as exhibited in the progression from light to heavy nuclei in the periodic table of the elements, is not so easily changed, even by nature. As nuclei become heavier with the addition of more neutrons and fewer protons, they become less stable against spontaneous disintegration until, in the heaviest stable nucleus, the introduction of one more neutron leads to violent instability and a new phenomenon, which we shall discuss briefly in the next chapter.

For the declining star, Sanduleak 202, all this nuclear physics meant that the star continued to fuse lighter elements into heavier ones, hydrogen into helium, helium into carbon and oxygen, and so forth, with a release of energy in each of those reactions until all the lighter elements in its core fused into iron. Then all fusion stopped. The reason, as we have just seen, was that the fusion of two iron nuclei would produce a heavier nucleus with a lower binding energy per constituent than iron. Consequently, energy would not be released as a by-product of the formation of the heavier nucleus. Fusion of iron or elements heavier than iron would be a losing proposition for the production of energy. Sanduleak 202 used up all of its useful nuclear fuel when its core became iron, and so its fate was sealed.

10 DOOM AND THE AFTERMATH

If we were able to see inside a star at the moment that all fusion stops, we would find it roughly organized in layers of different elements, iron in the core and less heavy ones at the larger radii. This is indicated schematically in Figure 5. The lighter material farthest from the core is only weakly burned in stars that become supernovae. In some stars of small mass, of less than about eight times the Sun's mass, the elements in the outer region burn briefly but with only a small output of energy. In the more massive supernovae, the temperature outside the core is too low to permit significant fusion in the outer layers. Even when fusion outside the core begins

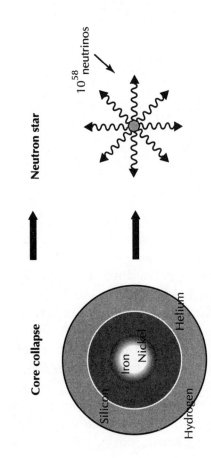

Core collapse → **Neutron star**

Silicon

Iron Nickel

Helium

Hydrogen

10^{58} neutrinos

FIGURE 5.   Illustration of the concentric layers of the chemical elements comprising the star Sanduleak 202 shortly before its iron and nickel core collapsed. With its collapse to a much smaller diameter, the enormously hot core was reduced to individual neutrons and protons. Instantaneously, protons and electrons combined in a "weak" reaction to produce more neutrons, and an incipient neutron star was formed from the core. Reducing the very high temperature of the incipient neutron star required the emission of an extraordinarily large number, 10^{58} —ten multiplied by itself fifty-eight times—of neutrinos.

and continues for a brief period, it does not save the star from destruction.

The star is now at the end of its life, an end that, with the implosive force of gravity, comes with astonishing rapidity and violence. At one instant the diameter of the core of the star is perhaps 10,000 or 20,000 kilometers. As gravity works its will, after a few tenths of a second the core undergoes a massive shrinking to roughly a few tens of kilometers in diameter. The iron in the outer layers of the core has fallen inward with unbelievable speed, indeed, at almost the speed of light. The iron falling in on the newly created, smaller, much denser core meets an incompressible mass, momentarily rebounds, and collides with the material still coming in. This collision generates a spherical shock that moves outward with the speed of sound, not light, and explodes the outer layers of the star. It was this explosion that gave rise to the brightened sky that Shelton and McNaught saw and is shown in Color Plate 1.

The core collapses in only a few moments, but the shock resulting from the turbulence of the collapse travels more slowly than the infalling iron nuclei. It is intriguing to realize that as a consequence, the outer layers of the star remain as they were, unaware of the cataclysm taking place in the interior until, much later, the shock reaches and excites them and blows them away.

We have not yet finished with the iron core. It is so dense and so hot that the iron nuclei come completely apart into their now free constituent protons and neutrons, because the energy of these constituents' motion in the overheated iron exceeds the total energy with which they were bound. The freed protons seize on free electrons—whose presence in the core we have, for simplicity's sake, so far neglected—in a reaction in which a proton and an electron coalesce to become a neutron and a neutrino. In an instant, all the protons and electrons are used up; the neutrinos soon depart; and the neutrons from the reaction join the other neutrons in the core to form an incipient neutron star, or what is sometimes called a *protoneutron star*.

An incipient neutron star is all that remains, because as a remnant of the original core it is still much too hot—10,000 times hotter than the core of the Sun—to be in a stable energy configuration. This enormous white hot mass of neutrons must make the transition from a state of high internal energy to a state of lower internal energy if it is to survive as a single stellar entity. To make the transition successfully and reach a stable energy configuration, it must rid itself of the vast amount of energy that is the difference between the higher- and lower-energy states. In brief, the star must find a way to cool itself. The failure to do so, and promptly, will cause many of the neutrons to boil away despite the gravitational force holding the incipient neutron star together, just as water molecules in a pot on a hot stove boil away despite the forces holding the liquid together.

Given sufficient time, the neutron star-to-be might dispose of the excess energy by radiating it away as visible light, X rays, and the like. But as we observed in the case of the Sun, electromagnetic radiation would remain trapped in the enormously dense mass of neutrons and in the blanket of the outer mantle of Sanduleak 202. The time required to jettison the excess energy would be thousands of millennia. But the neutron star-to-be cannot wait that long, as a significant fraction of it would vaporize into a gas of free neutrons that would be blown away by the stellar winds. Electromagnetic radiation is not the means by which the excess energy can be released.

Only one recourse is open to the star-to-be if it is to become stable and remain whole: it must dispose of its excess energy in the form of neutrinos. Neutrinos can penetrate a dense stellar core and escape from it in a few seconds, as they do when coming to Earth from the center of the Sun. There is, however, one important difference between the two escapes. The less than 3 seconds that it takes the neutrinos to travel the 700,000 kilometers from the center of the Sun is simply the travel time. But the 12 seconds during which neutrinos were emitted from SN 1987A was not the time to travel the few tens of kilometers across SN 1987A. That is, neutrinos were actually trapped by their many collisions in the remarkably dense core of SN 1987A—the only known environment in which neutrinos are trapped—and did not emerge directly from the center of the core but dribbled out from its surface during those 12 seconds. Moreover, the neutrinos serving to cool the neutron star-to-be do not come from nuclear reactions as do the neutrinos from the Sun. Rather, they emerge from the direct conversion of the star's excess energy to neutrino–antineutrino pairs, which then boil out of the incandescent neutron mass. This boiling away of neutrinos is the alternative to the boiling away of neutrons and the partial decomposition of the hot mass of neutrons. It is this process that allows the compacted neutron core to find stability as a neutron star.

The creation of neutrino–antineutrino pairs from the heat energy of a neutron star is a remarkable process in itself. We are used to the incandescence of hot objects, from which visible light is emitted. In those instances it is the excitation of the atoms in the hot matter—say, the filament of a lightbulb—that causes them to emit the low-energy photons that are visible light. Remember that photons belong to the class of elementary particles that are not subject to any limitation on their number, so that one or two or three or more can be created as an atom jettisons its excess energy. In contrast, neutrinos must satisfy the fundamental law that conserves their number. Therefore, the heat energy of the neutron star cannot release a single neutrino but must create a neutrino–antineutrino pair, each member of which has energy a million times greater than the energy of a photon from an incandescent object. Indeed, at its temperature of about 10^{11} degrees, the neutron star is one hundred million times hotter than the filament of a lightbulb. In all of this, the equivalence of mass and energy is exhibited again and again. These are phenomena occasionally witnessed in earth-bound laboratories, but never on the huge scale of mass and energy in supernovae.

Neutrino–antineutrino pairs cooled SN 1987A, and it was actually the antineutrinos from those pairs that the Kamioka and Lake Erie detectors observed. From the measured energies of the antineutrinos in Figure 4 and a well-known mathematical relation between energy and temperature, we estimated that the temperature of SN 1987A was originally about 10,000 times hotter than the core of the Sun. Furthermore, the total energy carried away from SN 1987A by the neutrino–antineutrino pairs corresponded to a mass approximately equal to 1.4 times the mass of the Sun. In this way, during the brief period of our observation, we estimated the power output of SN 1987A to have been the enormous total of 10^{39} megawatts that was mentioned earlier.

It is difficult for us to appreciate all the inferences to be drawn from these numbers. They convey how extreme the final compaction of Sanduleak 202's core was and how huge the amount of energy it needed to expel so that it could become a stable neutron star. But the numbers do not directly reflect the urgency of the star's need to be cooled. An analogous situation on a human scale might be helpful. Think of a person with a fever. If the fever is severe, in the vicinity of 105°F, quick measures are urgently required to carry away the body heat and reduce the temperature so as to prevent serious damage to the person. So it was with the collapsed, feverish core of Sanduleak 202. If it was to survive as a star—true, a burned-out star, perhaps ultimately a pulsar—it needed the quick relief of neutrino cooling to carry away the heat that threatened to burn it up.

This is a good place to compare the concise style of reporting usually preferred by scientists writing for one another with my somewhat more graphic style here. Some of my physicist friends would instinctively moderate my description of how the core of SN 1987A lowered its high temperature. It is certainly correct, they would say, that prompt cooling by photon emission was not readily available to the protoneutron star and that neutrino emission was. They would argue, however, that the central feature of such an event becomes apparent when we recognize that—in general, all else being equal—the emission of photons from an excited physical system is much more probable than the emission of neutrinos, especially neutrinos in neutrino–antineutrino pairs; that it was the remarkably high density of the excited neutron core that effectively canceled out the overwhelming probability of photon emission and led to the resultant neutrino emission. Nature simply acted, they would insist, in accord with one of its basic precepts: whatever is not expressly forbidden—no matter how improbable—will take place only if it is more probable in a given situation than the alternatives to it. Consequently, their argument would conclude, a simple recitation of the physical factors in the cooling of the protoneutron star gives a clear enough account of the marvelous subtlety of nature at work.

These differences in presentation serve a useful purpose because each emphasizes different aspects of the core's collapse and, indeed, of other issues as well. We stressed the speed and violence of the core's collapse, but we should be equally aware of the interplay between the fate of Sanduleak 202 and the physical laws that were manifested, first, as we saw, in the creation of the iron core and then in the method that nature chose to cool it. But there is more. We asserted earlier that antineutrinos, not neutrinos, initiated the reactions observed in the underground detectors, even though SN 1987A emitted both antineutrinos and neutrinos together. We cannot provide direct confirmation of that assertion from measurements made in the Kamioka and Lake Erie detectors. However, experiments at accelerator laboratories have conclusively confirmed that low-energy antineutrinos may interact with the essentially free protons in the detector water (two protons and an oxygen nucleus) but that neutrinos of the same energy cannot. This result is dictated by the law that electric charge and particle–antiparticle number must be conserved in the antineutrino and neutrino interactions. Furthermore, neither antineutrinos nor neutrinos with the low energies they possessed when emerging from SN 1987A were able to react directly with the oxygen nuclei in the water. Finally, the likelihood that an antineutrino or a neutrino would collide directly with one of the atomic electrons in the water is known to be about 10 times too small to account for the 20 observed events. Hence we concluded that antineutrinos, not neutrinos, initiated the observed events and that the charged particles actually observed were positively charged positrons, the antiparticles of electrons. The same reasons dictated that the earliest burst of neutrinos from the core of SN1987A—the neutrinos arising from the reaction in which the freed protons from the disintegrated iron nuclei were converted to neutrons by combining with free electrons in the core—did not produce the events observed by the two detectors.

A related question we asked was whether it was possible to use the directions of the observed positrons to point their parent antineutrinos back to SN 1987A. This was the method used by the Kamioka collaboration to detect the electrons produced by solar neutrinos and to point those neutrinos back to the Sun. Could we also do so for the antineutrinos from SN 1987A in a similarly convincing way? How exciting it would be to find—from underground, even—the source of those antineutrinos in the sky 160,000 light-years away! It turns out that for good reason, we could not apply this method to the antineutrinos from SN 1987A. The only reaction that allows the direction of the incident antineutrinos to be determined is the one in which the antineutrino target is an atomic electron with its very small mass that moves from the collision in the same direction as the antineutrino. But we have just pointed out the low probability of that reaction—10 times smaller than the probability of antineutrinos' finding protons as their targets. When protons are the targets, the connection between the direction of the incoming antineutrino

DOOM AND THE AFTERMATH

and that of the outgoing positron is lost because the proton is almost 2000 times heavier than the electron. Accordingly, we expected that the observed positrons, although certainly produced by antineutrinos from SN 1987A, would not show a directional correlation with it, and indeed no convincing one was found.

The situation is different with respect to the directional detection of neutrinos from the Sun. Only neutrinos, not antineutrinos, are created in nuclear fusion reactions. Consequently, only neutrinos emerge from the Sun, since it is not nearly hot enough to boil away neutrino–antineutrino pairs. Again, we have just seen that low-energy neutrinos cannot interact with the hydrogen and oxygen nuclei in the Kamioka detector water. Therefore, the less probable but still possible reaction of neutrinos with atomic electron targets dominated the action of solar neutrinos in water and, at Kamioka, allowed the solar neutrinos to be pointed back to the Sun.

We also considered the importance of nuclear fusion in the life of Sanduleak 202 and the absence of nuclear fission, the other nuclear phenomenon in which vast amounts of energy are released. We wondered at this absence in view of the fact that nuclear fission is currently our principal method of generating nuclear energy. Remember Sanduleak 202's attempt to survive. It fused all its core nuclei until the core was solid iron. Fusing iron to nuclei of heavier elements, we said, produces no energy from the star because the binding energy per constituent steadily decreases in the progression from iron to uranium, the heaviest stable element.

The nuclei of those heavy elements have been explored in laboratories in Europe and the United States since the early 1930s. It came as a stunning surprise to the physicists and nuclear chemists doing this work when they realized in 1938 that one of the natural forms of uranium teetered on the brink of instability because of its excess of neutrons. This was not a known mode of instability but a new, previously unrecognized, much more violent mode. This new phenomenon—the fission or splitting of uranium—is shown schematically in the graph of binding energy per constituent plotted in Figure 6 against the number of constituents for all nuclei, from the lightest to the heaviest. Starting at the far left in Figure 6, one is in the region of the elements used as fuel by SN 1987A, the region in which fusion takes place. The binding energy per constituent increases as we move along the curve in Figure 6,

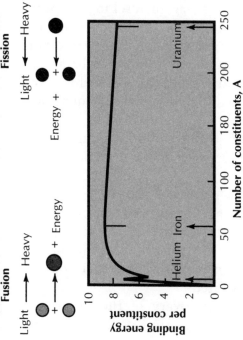

FIGURE 6.   Curve showing the dependence of the nuclear binding energy per nuclear constituent on the number of constituents (A). The constituents are neutrons and protons. The peak at A = 4 corresponds to the exceptionally stable helium nucleus. The binding energy per constituent has its maximum value in the stable nucleus iron with constituent number A = 56. The fusion of light nuclei to form a heavier one releases energy, as shown in the small diagram at the top left. The fission of a heavy nucleus into lighter ones also releases energy, as shown in the small diagram at the top right. Once iron is reached, however, it's all "downhill" for energy generation by either fusion or fission. Adapted from Arthur Beiser, *Concepts in Modern Physics*, 4th ed., McGraw-Hill, New York (1987).

rapidly at first and then more slowly as it reaches elements in the vicinity of iron.

Alternatively, starting at the far right in Figure 6 at uranium and moving to the left along the curve, we see that the binding energy per constituent also increases, again reaching its peak value at iron. As a consequence, energy could be generated if a reaction were found that did not require much input energy and would break up a heavy element into lighter ones. In this way, energy initially residing in the loosely bound, almost unstable heavy nucleus would be released in the formation of the more tightly bound products of the breakup of the heavy nucleus. According to Figure 6, energy is released in fission by starting at the right and moving left along the binding energy per constituent curve, and energy is released in fusion by starting at the left and moving to the right. In each case, the energy stored in nuclei or in the process of their formation is what is liberated.

The accidental discovery of the fission of one form of uranium shortly before World War II alerted the scientific community in Europe and the United States to the possibility of an "atomic" bomb. In a dramatic sequence of events, the American and British governments were led—largely by distinguished émigré European scientists, foremost among whom was Albert Einstein—to consider this seemingly remote possibility as a potential determinant of the course of the forthcoming war. Germany and Japan also recognized this possibility but could not act on it so quickly. The principal criterion for a bomb was that the fission of each uranium nucleus provide extra neutrons to induce fission in nearby uranium nuclei, thereby sustaining a chain of successive fission reactions in a mass of uranium. This criterion was shown to be satisfied in the operation of the first nuclear "reactor," constructed by a group of young scientists under the direction of the ubiquitous Enrico Fermi at the University of Chicago. The creation of the first atom bombs at the Los Alamos National Laboratory in New Mexico followed, under the direction of J. Robert Oppenheimer. That story has been told many times in the half-century since the bombs' first and only use in a war.

Figure 6 says it all, describing dispassionately the principle entailed in the generation of these huge amounts of nuclear energy. As far as we know, nature does not make important use of fission in astrophysical or other phenomena, even though roughly 10 times more energy is released in a single fission than in the fusion of four hydrogens to helium. Rather, as the stars tell us, fusion is the preferred means of large-scale energy generation in nature, presumably because hydrogen is so abundant and the heavy elements are so rare in the universe. Some day, we too may use fusion, though on a much smaller scale than in a star, to replace our present energy-producing plants based on fission.

In the broadest sense, the events comprising SN 1987A are not mysterious but are well understood. The progression of events leading to the collapse of the star's core and the aftermath of neutrino cooling conform to the laws of physics, with little room for speculation and certainly none for an explanation based on superstition. Nevertheless, some issues remain uncertain. For example, we noted that the outer regions of the star were excited and blown away by the shock that propagated outward from the pulsing core of Sanduleak 202. We must admit, however, that a detailed, quantitative description of the violent ebb and flow of such a huge amount of matter is almost beyond us. It is probable that gas in the region between the collapsed core and the shock is heated by the vast number of neutrinos and explodes the outer part of the star. We know that the energy carried by the neutrino–antineutrino pairs is 100 times greater than the energy in the shock. This leads us to point to neutrino heating as the mechanism to boost the shock past the stall, as suggested by mathematical models of core collapse and shock propagation. Nevertheless, we still need to round out our understanding of the supernova phenomenon in this and other regards.

11 TWO IMPORTANT CONSEQUENCES OF SUPERNOVAE

The temperature at the center of the Sun is high enough to fuse four hydrogens into helium in the principal fusion reaction. It is just about high enough to enable other fusion reactions, which lead to the formation of small quantities of elements with more constituents than helium, for example, lithium with three protons and four neutrons, beryllium with four protons and five neutrons, and boron with five protons and six neutrons. From studies in nuclear physics laboratories, we can calculate quite accurately how much helium and how much of the heavier lithium, beryllium, and boron are produced by the fusion reactions in the Sun. And we can use the Big Bang theory to estimate the large

quantities of hydrogen, helium, and lithium produced just after the Big Bang when newly formed
protons and neutrons were free to combine and
before all the free neutrons had decayed away.
When we compare these estimates with the relative quantities of the same elements found elsewhere in nature—on our Earth, in other stars, and
outside our galaxy—the agreement is good. We
believe that the creation of the lighter, less complex elements in the early universe—the process
known as *nucleosynthesis*—took place primarily
through combinations of the very hot matter emanating from the Big Bang and, to a lesser extent,
by fusion in the stars. The success of the theory
of nucleosynthesis is one of the cornerstones of
the Big Bang cosmology.

For a long while, however, twentieth-century
astronomers and cosmologists were at a loss to
explain the existence of elements with more
nuclear constituents than the elements created by the Big Bang or fusion in the stars. The
temperature soon after the Big Bang and at the
center of stars such as the Sun was too low to
fuse the lighter mass elements into heavier elements with more constituents. Where, we wondered, was there a furnace capable of reaching
the temperatures necessary to create the more
complex elements that we see around us: the

silicon in glass, the iron in steel, and the gold in our jewelry?

By now, you know the answer to that question: it was the core of a collapsing star. This idea was first proposed by the astronomer Fritz Zwicky who, in the second quarter of the twentieth century, called attention to the importance of supernovae. His research led him to conclude that supernovae were the only furnaces generating sufficient heat to manufacture the more complex elements. Zwicky made this suggestion well before we knew much about supernovae and even before we had constructed our current model of the Sun. Incidentally, he also speculated—soon after Wolfgang Pauli suggested the existence of neutrinos, as did George Gamow and M. Schönberg, and almost half a century before SN 1987A—that neutrinos would be emitted from supernovae.

In its core, in an effort to survive, a dying star produces quite complex elements. After its core collapses, the very hot incipient neutron star also makes available a large supply of free neutrons. These are required as the constituents of the neutron-rich stable elements heavier than iron. Zwicky maintained that the countless supernovae of the past spewed into space the matter out of which was built the chain of elements present in the periodic table. This chain is limited in length by the instability of its heaviest link, which is destroyed by fission when one more neutron is added to it.

We have no definitive proof that supernovae are the principal—possibly the only—source in nature of the heavy elements, but we are not aware of any other phenomena in which the conditions of temperature, density, and pressure can lead to the fusion of the elements up to iron and to the release of such huge numbers of free neutrons. The conjecture that supernovae might be the manufacturers and distributors of the elements up to iron, and of the neutrons that build the heavier elements, was verified through the observation of the burst of antineutrinos from the collapse of the iron core of SN 1987A.

With supernovae as an integral step in this process, we have formed a rudimentary blueprint of the origin of the elements. As with all origins, the beginning is hidden in the mists, in the condensation of quarks that are the elementary constituents of neutrons and protons formed after the Big Bang. Whence came the quarks and the other fundamental constituent of matter—the electrons—we do not know. The lighter-mass elements were formed as neutrons and protons combined soon after the Big Bang and in the furnaces of newly created stars. As the stars aged, the supernovae that heralded their deaths produced nuclei in the midrange of nuclear masses and freed neutrons to be captured and to form the heavier elements. The periodic table of the elements does not, however, progress to heavier and heavier elements but is terminated by the increasing instability of the neutron-rich heavier nuclei. We should marvel that our place—even our existence—in the universe is dependent on the detailed functioning of stars. Yet without sodium, without potassium, without iodine, life on Earth as we know it would not exist. It is one of the attractions of science that as we learn more about the universe, we learn more about the circumstances that contribute to our survival in it.

Many other inferences of scientific importance have been drawn from the observations of SN 1987A, but only a few of them have had a cosmic impact comparable to the one we just discussed. One of them, however, we will describe here because it impinges in a simple and yet profound way on a very abstract idea that fascinates scientists and nonscientists alike. It, too, is structured on a grand scale and makes us appreciate even more the subtlety of the universe.

Some 11 years after publication of his special theory of relativity in 1905, Einstein developed a theory of space and time that was more general than the special theory. This later theory became known as the *general theory of relativity*, although its main focus, unlike that of the special theory, is gravitation, the universal attractive force between any two masses and the resultant accelerated motion produced by that force. Subsequently, Einstein accepted the challenge to construct a theory that unified gravitation with the other fundamental forces in nature, a challenge that occupied him for the rest of his life. The general theory, like the special theory, has revolutionized our thinking about the universe, despite Einstein's failure to find an ultimate unifying theory. For that matter, no one else has succeeded in constructing such a theory.

The special theory emphasizes phenomena in which bodies move with constant velocities relative to one another, and it elaborates on the behavior of those phenomena as the velocities approach in magnitude the velocity of light. An axiom of the special theory is that the velocity of light is the maximum permissible velocity in nature, which can be achieved only by particles without mass and can only be approached but never attained by particles with mass. A fundamental consequence of this axiom is that the *time* dimension assumes parity with the three space dimensions and that relative time becomes as meaningful as relative distance and relative velocity.

The general theory is a theory of space and of the warping or distortion of space by the presence of mass. It is a theory involving curved space, not the flat space of Euclid that has dominated our geometry for two and a half millennia. The curvature of space reflects the presence of mass and the attraction of every massive object to every other massive object. The general theory is even today a frontier of research in astrophysics and cosmology, an incomplete legacy from Einstein's towering intellect. In both the general and the special theory, mass and energy are again interchangeable, and it is no surprise that light is acted on by the gravitational force or, stated in another way, the trajectories of photons are deflected by the gravitational action of mass, even though the photons themselves are without mass. As a consequence, light traveling a great distance between two points in space moves along a slightly curved path and takes a longer time to cover that distance than it would if it moved along a straight path between the two points. According to the general theory, this curvature of path is due to the masses of stellar bodies that act on the light through the gravitational force they exert. Because gravity is a long-range force, the galaxies that bend the path need not be along or even especially close to the path; the actual bending is produced by the weighted average distribution of mass "felt" by the light over the distance it travels.

For distances in everyday life, and even those in our solar system, the deflection of light is at the limit of our measurement capability. But according to the general theory, the path of the visible light coming from the excited mantle of Sanduleak 202 in its trip to Earth is curved by the effect of the scattered masses in the Milky Way. This creates a significant time difference between travel along this curved path and travel along a straight path from Sanduleak 202 to Earth. This difference has been estimated to be about 5 months. Relative to the approximately 160,000 years that a trip along either path requires, 5 months is a short time indeed, but still of interest. We don't know either the distance to Sanduleak 202 or the magnitude of the dispersed masses along the path from Sanduleak 202 to Earth well enough to make a precise comparison of the time difference between travel along the curved and the straight path. So we cannot directly test the bending of light predicted by the general theory. But we can use the emission of both neutrinos and photons from SN 1987A to test whether the general theory applies equally to all types of particles. According to the general theory, the gravitational force should act on neutrinos in the same way as it acts on photons. If it does, neutrinos and photons should travel along the same paths, whether curved or straight, and therefore the neutrinos and the photons from SN 1987A should have reached Earth at the same time, allowing for the difference in the different times of emission from SN 1987A.

Earlier we talked about the characteristics of the two classes of elementary particles to which photons and neutrinos belong. We emphasized the difference between them arising from the fundamental constraint on, or conservation of, the total number of neutrinos and the absence of that constraint on photons. If the general theory is truly a universal theory, it must apply with equal validity to the two classes of elementary particles, to neutrinos as well as to the photons, for which the theory was originally developed. The emission of both neutrinos and photons from SN 1987A provided a good test of that aspect of the theory, an aspect important enough to be known by the impressive title *the weak equivalence principle of the general theory of relativity.* Among other things, the principle states: "Any uncharged body traveling in space will follow a trajectory independent of its internal structure and composition." This requires that the two "bodies," the neutrinos and the photons—although presumably different in "internal structure and composition"—should have followed the same trajectory and consequently reached Earth at the same time.

We mentioned earlier that the detection of neutrinos preceded the first visual sighting of SN 1987A by 3 hours at most. For want of a more precise measurement, we took 3 hours as the maximum interval between the time taken by the neutrinos and the time taken by the photons to travel to Earth from SN 1987A. We then compared that 3-hour interval with the 5-month interval that it would have taken if the neutrinos had traversed the straight path while the photons traveled the curved path. Dividing 3 hours (10,800 seconds) by 5 months (1.35×10^7 seconds), we found that the neutrinos must have traveled to Earth along precisely the same curved path taken by the photons, within an uncertainty in time of less than one-tenth of 1 percent, the uncertainty being our not knowing the exact time the photons were emitted. The weak equivalence principle passed this test, the first comparison of the space travel of neutrinos and photons. This result was first obtained independently by Michael J. Longo and by Lawrence M. Krauss and Scott Tremaine.

Einstein's general theory is an extension of Newtonian theory, and principles such as the weak equivalence principle are extensions of the fundamental laws of motion of Newtonian mechanics. It is likely that the equivalence principles of the general theory—the weak one just discussed and a stronger one not relevant here—will be taught to elementary physics students in the centuries to come, just as Newton's three principles of motion have been for centuries.

One more result of interest was obtained from the 3-hour upper limit on the time interval between the detection of neutrinos and photons. We compared the velocity of neutrinos with the velocity of light to see how nearly equal they appeared to be. We found that the velocities of neutrinos and photons traveling from SN 1987A to Earth differed by only six-tenths (0.6) of a meter per second out of three hundred million (3×10^8) meters per second, only slightly larger than the uncertainty in the actual measurement of the velocity of light itself. Finally, the spread in time of the individual events recorded in the Kamioka and Lake Erie detectors (see Figure 4) led us and many others to the limiting value of neutrino mass that we used in Chapter 4. Some day, a supernova explosion closer to Earth than the Large Magellanic Cloud will reveal a much larger number of observed neutrino events in neutrino telescopes built especially for that purpose. It should be possible from the time structure of events in such a neutrino burst to determine a limiting value of the neutrino mass far superior to the one we now have.

This question of the masses of the elementary particles is one of the most exasperating in particle physics. The masses extend from the very small—possibly even zero—value of the neutrino to that of the heaviest quark almost a million times larger than the mass of the electron. This wide range is without rhyme or reason, which is to say that we have no model, no theory of the elementary particles in which the particular values of their masses can be explained in a convincing way. To make calculations concerning the elementary particles, which we do all the time, we put into the equations by hand the measured values of the masses acquired through experiment.

This is better than nothing, but it leaves us feeling dissatisfied and insecure.

A final word. To understand what stirs scientists' inner beings, try to appreciate new discoveries that illustrate the inherent order in the universe and new tests of principle of the kind in this chapter in the same way that you would a newly found poem or a newly heard concerto that momentarily lifts you out of yourself. Far removed as they are from the pragmatic reality of our daily life, the ideas we have discussed—elegant and impersonal—are the essence of science. Through each of them we elicit a bit of the basic structure of the marvelously complex universe in which we are so small a part.

If this were strictly a scientific monograph on SN 1987A, we would go on here with more of the same, but it is not. Instead, we shall look at what the first years of life have brought to the remnant of SN 1987A and so move our story to its end.

12 NEUTRON STARS, THE HUBBLE SPACE TELESCOPE, AND BLACK HOLES

Ten years have passed since the burst of neutrinos from SN 1987A. What have we learned about the exploded star in that time and, in particular, about the compact mass at its center? Are we sure, as we have assumed, that there is a neutron star hidden in the remnant of Sanduleak 202? The answers to these questions are not easily found, despite all the study of SN 1987A in the interim.

First, neutron stars were recognized as distinctive stellar entities only as late as 1968 with the observation by England's Anthony Hewish and Jocelyn Bell of objects now known as *pulsars*. This recognition was slow in coming, even though Walter Baade and Fritz Zwicky, both in the United States, had advanced the idea as early as 1934 "that supernovae represent transitions from ordinary stars into neutron stars, which in their final stages consist of extremely closely packed neutrons." This statement testifies to their foresight as well as to their alertness, as neutrons had been discovered by James Chadwick in England only two years earlier.

Later in 1968, the United States' Thomas Gold proposed the now accepted interpretation of the Hewish and Bell observation, that pulsars are rotating neutron stars. Pulsars are primarily radio sources, and they are observed with radio telescopes similar to those in Color Plate 3. As their name suggests, pulsars periodically pulsate at high frequencies, from several to many pulses per second. Gold made the connection between a rapidly rotating neutron star that had a strong magnetic field and a pulsar's intense narrow beam in the radio band that was observed to sweep regularly by a radio telescope's line of sight. Only a neutron star is likely to be born while rotating that rapidly—a consequence of the conservation of angular momentum—and only a neutron star is bound together tightly enough to spin at those high frequencies without coming apart. Finally, in this model of the pulsar as a rotating neutron star, the source of any pulsed but generally weaker optical signals would be the strong magnetic field trapped in the core of a progenitor star as it collapsed into a neutron star.

The point of this brief excursion into the subject of pulsars is to demonstrate that the seemingly burned-out neutron star can and does make its presence felt. What does this mean for SN 1987A? For a short time, it was thought that a weak optical pulsar, rotating at an extraordinarily high frequency, had been detected in the supernova remnant, but this turned out to be wrong. And as yet there has been no positive radio signal from the remnant, although many radio telescopes continue to observe it carefully. It would be important to identify a pulsating neutron star in the center of the remnant and thus begin a detailed observation of a pulsar so soon after its birth. But there are several good reasons that a magnetically active neutron star may actually be there but cannot be observed right now. We shall just have to wait.

Modern astronomy is multifaceted, as we have already seen. Since SN 1987A, another facet was discovered by the Hubble Space Telescope—an optical telescope mounted on an earth-orbiting satellite and named for the astronomer Edwin Hubble, who first suggested in 1928 from his optical observations that we live in an expanding universe. After the crew of the space shuttle *Endeavor* corrected an aberration in the lens, the telescope now takes beautiful, high-resolution pictures of stellar objects, undistorted by the Earth's atmosphere, which the satellite transmits electronically to Earth. These observations have realized the dream of those who conceived and brought this project to fruition. The Hubble telescope has influenced our view of SN 1987A, and consequently its observations deserve mention in this story.

To complete our catalog of telescopes, Color Plate 10 shows the Hubble Space Telescope immediately after it was released from the shuttle's cargo bay. To point the telescope at a given object in the sky and maintain it in position while pictures are being taken is a demanding task. It is necessary to know the exact location and velocity of the satellite and to have continuously available from its onboard computer the precise coordinates of the object under study. This impressive accomplishment has become routine as the telescope photographs one stellar body after another and sends its photographs (actually, digitized electronic data) to the Space Telescope Science Institute at the Johns Hopkins University in Baltimore, Maryland.

Color Plate 11, taken in August 1990, shows an elliptical ring of gaseous material around the remnant of SN 1987A. The tightly bunched debris, the remnant of the stellar explosion, is the red blob near the center of the ring. The image of SN 1987A grows in size as the remnant continues to expand. The blue stars on either side of the ring are not associated with the supernova. From their size and the size of the stars near SN 1987A in Color Plate 1, we can get a rough idea of the better resolution in Color Plate 11 compared with that in Color Plate 1, even though the angular scales of the two figures are different. The red and blue colors in Color Plate 11 reflect the relative colors of the supernova remnant and the nearby stars.

It was thought that the yellow ring would last for perhaps a century, but not much longer than that because by that time the debris expanding from the remnant would overtake and disintegrate the ring. Imagine the surprise of the astronomers when they received the photograph in Color Plate 12, also taken by the Hubble telescope, but in May 1994. Now three elliptical rings of glowing gas encircle the remnant of SN 1987A, the original ring that is seen in Color Plate 11, and two larger ones. Although all the rings in Color Plate 11 appear to intersect, it is unlikely that they do so in real space. The small bright ring lies in the plane containing SN 1987A, and the two larger rings probably lie in front and behind.

These marvelous photographs seem to tell us very little about what is hidden in the remnant of SN 1987A or the space immediately nearby, but a possible interpretation of the two larger rings, advanced by Christopher Burrows, a member of the European Space Agency/Space Telescope Science Institute, suggests otherwise. Burrows thinks that the rings are the result of a high-energy beam of radiation sweeping across and exciting an hourglass-shaped bubble of gas blown into space by the original supernova explosion. There is evidence for the gas bubble from earlier observations. According to Burrows's interpretation, the geometry of Color Plate 12 may actually be as sketched in Figure 7, which, if correct and accurate, indicates that the source of the radiation responsible for the excitation of the gas envelope is more likely to be a previously unseen stellar object (itself a remnant) that was a binary companion to Sanduleak 202, rather than an object within the remnant of SN 1987A.

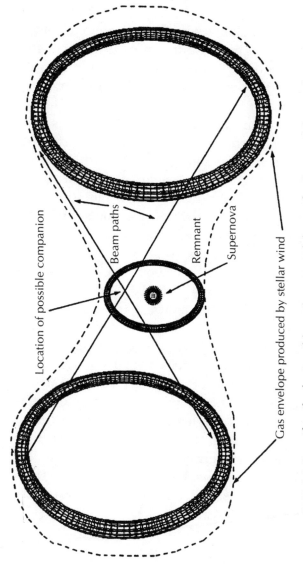

FIGURE 7. Sketch of a possible explanation of the configuration of the two larger gaseous rings in Color Plate 12. The two rings are thought to have been produced by a beaming electromagnetic source located on an unseen companion of the SN 1987A remnant, acting on a previously detected gas envelope.

Location of possible companion

Beam paths

Remnant

Supernova

Gas envelope produced by stellar wind

The rapid changes in the environment of SN 1987A shown by the differences in Color Plates 11 and 12 highlight the importance of frequently revisiting that remnant with the Hubble Space Telescope during the next decade. The future behavior of the radiation-excited gas rings may strengthen Burrows's interpretation and may indicate the source of radiation to be within the SN 1987A remnant. In any event, if confirmed, the presence of a beaming source of radiation will serve as a probe of the space in or around the remnant and help identify what remains of the explosion of Sanduleak 202.

The lack of certainty about the object at the center of Color Plates 11 and 12 opens the door to the possibility that contrary to what we have assumed all along, the central object might not be a neutron star. An alternative would be an object originating, as does a neutron star, from the collapsed core of a dying star, but one so massive and compact that it is able to carry the properties of the neutron star to their limit. As evidence for its existence has mounted, this object—appropriately named a *black hole* by John Wheeler in 1968 and once a theoretical curiosity—has become a focus of interest to scientists and nonscientists alike.

If you were to ask a theoretical physicist to define a black hole, he or she might absentmindedly tell you that it is the name given to a singularity in the solution of certain equations in Einstein's general theory of relativity and that in addition, a singular solution is one that tends numerically toward infinity as some specified quantity tends toward zero. In view of how far we have come in the last few chapters, this response might not be totally without meaning, but it still lacks a model to help you visualize what is being talked about.

The request to be a little more concrete might produce a different approach. Visualize, you might be told, the collapsing iron core of a supernova, and ask yourself: what stops the collapse? Remember that gravity, the attractive force of the core's innermost mass, is acting from all directions on the core's outer regions and is sucking them inward radially, as if there were a vacuum at the star's center. As more and more mass falls in, the force exerted on the outer regions increases because the force is directly proportional to the amount of mass at the center. So again, ask yourself: what stops the collapse?

The conventional answer—which we offered earlier—is that in the very hot core, the neutrons and protons from the already disassociated iron nuclei form a kind of incompressible fluid when the distance between them reaches a certain small value. The infalling outer layers of the core pile up on this extremely dense fluid and also become extremely dense. But—the conventional answer asserts—the radius of this inner region does not become less than the value allowed by the incompressibility of nuclear matter.

This answer rests on the important assumption, among others, that the collapse occurs smoothly despite its speed, that the mass builds up at the center slowly enough to allow the iron nuclei to be stripped down to their constituents, and the constituents to repel one another, so as to reach the transient equilibrium state described as a white hot, throbbing mass of neutrons. But also remember how many things are going on at essentially the same time as the collapse— the conversion of protons to neutrons and the initial emission of neutrinos from that process, the generation of a series of sounds and shocks and the enormous numbers of ions and electrons they produce, and, finally, the emission of 10^{58} neutrinos to cool the collapsed core.

Suppose that because of all that turmoil, the infalling matter overshoots its target, as often happens in nature, and passes through the state dictated by nuclear incompressibility before equilibrium in that state can be established. Then, at least conceptually, there would be no obvious limit to the compressibility, no limit to the allowable density of the matter, and mass would pile on mass in a smaller and smaller volume until nature realizes the singular solution that our theoretical physicist mentioned. More to the point, once the collapsed core settles down to a deeper equilibrium in its superdense state, nothing would be able to leave it, no massive particle, no light, no neutrinos. Its gravitational attraction would hold fast its own constituents and entrap all other objects as well. The velocity that any object would need to escape this gravitational attraction would exceed the maximum allowable value, the velocity of light. The superdense state would thus act as a giant maw—in short, it would be a black hole.

The ostensible properties of black holes as they are hypothesized today are simple. The space around a black hole is bounded by a surface, the so-called horizon, beyond which there is no communication. The radius of a spherical horizon, named after Karl Schwarzschild, who first proposed the black hole solution of the Einstein equations, can be calculated. It leads to the conclusion that, for example, a stellar body with the mass of the Sun would be a black hole if it were compressed into a sphere with a radius of 2.5 kilometers, almost 300,000 times smaller than the actual solar radius of 700,000 kilometers. This in turn means that the density (mass divided by volume) of the black hole would be about 10^{16} times greater than the average density of the Sun and about 20 times greater than the density of a neutron star.

The density of a neutron star is approximately equal to the density of the nuclei in the light atoms. But most of the volume of an atom is empty, and atoms can't be packed as closely as can the neutrons in a neutron star. Consequently, the matter of the neutron star is different from any matter with which we are acquainted. To the best of our knowledge, however, the quark structure of protons and neutrons mentioned earlier is still valid at that density. Is it also valid at the singularity that represents the black hole? We don't know! Nor do we know if the matter of a black hole resembles any matter familiar to us! We are unable to simulate, much less reproduce, such matter in our scientific speculations, and certainly not in the laboratory. It may represent a natural extrapolation of the matter we understand or, equally likely, may be a new form casting new light on elementary particles and the structure of matter.

What is clear is that the formation of black holes through the supernova mechanism of a collapsing stellar core is a realistic alternative to the formation of neutron stars. It is not the product of a science fiction writer's imagination. The properties of black holes and neutron stars are probably both qualitatively and quantitatively different, and they may differ also by the introduction of scientific phenomena or principles beyond those already established. In brief, then, the formation and existence of black holes may seem outlandish, but there is no reason that they should be especially rare or be considered unnatural. Rather, the problem they create for us is how to detect them, which is not all that different from the problem of detecting neutron stars hidden in the remnants of supernovae.

Nevertheless, there may be differences between neutron stars and black holes that could help us distinguish between them. For example, although the neutron star's pulsating property occurs often in nature, we have not observed this property in black holes. At present, we envision black holes as ideally spherical, having no structure on which to attach a source of pulsating radio emission, but it is possible for a black hole to rotate and violate that restriction. If, then, the compact object at the center of the remnant in Color Plates 11 and 12 is ultimately observed to be a pulsar, it is more likely to be a neutron star, although a black hole cannot be ruled out conclusively.

Alternatively, supernovae occasionally occur near another star that survives the explosion and continues to be visible. This new binary system, consisting of a neutron star or a black hole and a normal star, may provide the means to distinguish between a neutron star and a black hole. In time, the normal star may expand and shed matter onto its invisible companion. This matter, ejected at different angles, will tend to spiral onto the companion, owing to the strong gravitational attraction of the companion—just as water spirals down a drain—and to form what is known to astronomers as an *accretion disk*— a spinning mass of hot gas. But the motion of the gas disk is not the smooth, laminar flow of water in a bathtub; it is the turbulent, chaotic motion of highly accelerated gas molecules at a very high temperature, in frequent collision with one another and with the center of gravitational attraction, the surface of the neutron star, or the event radius of the black hole. The result is likely to be a copious emission of X rays in the case of the neutron star and possibly of more energetic X rays or even gamma rays in the case of the black hole, because the infalling matter has farther to fall to reach the event radius of the black hole. In either case, the emission of X rays or gamma rays requires the gas to have a particularly high excitation energy, which we cannot explain unless we attribute it to the presence of the invisible companion, the neutron star or black hole. At least one such strong source of X radiation, Cygnus X-1, is thought to be a candidate for a black hole, mainly because it appears to be too massive to be a neutron star. Clearly, we need more observation of the star by satellites equipped with energetic radiation detectors, as well as observation of other potential candidates, before we can confirm this as a black hole.

CONCLUSION

The first modern optical observation of a supernova was made by E. Hartwig at the Dorpat Observatory in Estonia, on August 20, 1885. That supernova was near the nucleus of the Andromeda Nebula, but it was never identified as to type or detected with other modern instruments, in part because its remnant diminished extraordinarily quickly. Until SN 1987A, a century later, no other supernova had been observed so close to the Earth, and none has been visible to the naked eye since Kepler's supernova in 1604.

Nearly a thousand extragalactic supernovae have been found and cataloged since Hartwig's observation, as interest in supernovae has mounted over the years. But only a very few of them have been studied. It is no wonder, then, that the announcement of the sighting of SN 1987A galvanized astronomers into excited activity. Added to the visual observation of the exploded star was the detection of antineutrinos emanating from it, crystallizing 50 years of speculation and hypotheses about supernovae. Furthermore, the detection of antineutrinos from SN 1987A, in a galaxy beyond our own, and the measurement of neutrinos from our own Sun firmly established the new science of neutrino astronomy. At the same time, it has led to new efforts by experimental and theoretical physicists to determine the intrinsic properties of neutrinos themselves.

It was my purpose to convey the flavor of this scientific activity in recounting the story of SN 1987A. If I have succeeded, you should have learned about the development of a new type of telescope and the difficulties encountered in making it carry out the task planned for it. Perhaps also, you will now understand the supernova phenomenon and how it is governed by the laws of physics. You will have seen that the inferences drawn from observation of the different emissions—photons and neutrinos—from SN 1987A extend beyond the observation itself. And you should now be aware that the compact object residing in the remnant of SN 1987A may still have many important things to teach us.

Over and above the science are the human aspects of the scientific research: the vision that led to a concentration of effort in what turned out to be an immensely profitable direction, the discipline of the scientific method, the element of luck, and the ability to take advantage of a fortunate turn of events. If you have recognized in this book these aspects, as well as the science, it has indeed served its purpose.

The sighting and study of SN 1987 constitutes what we defined at the beginning of the book as a single work of science, single because the many diverse features of a rare occurrence in nature, far distant from us in space and time, were unified by our understanding of it; and single because such an occurrence serves as one of many illustrations of the natural order to be found in the universe.

EPILOGUE: NEUTRINOS FROM THE SUN

It seems a shame to have come this far and not to pause long enough to describe the outcome of the Kamioka detector's real-time, directional measurement of neutrinos from the Sun. It was the prospect of making that measurement that led to the modifications of the detector described earlier and that, with the help of a great stroke of luck, made possible our observation of the neutrino burst from SN 1987A. Soon after the excitement of that event subsided, our acquisition of solar neutrino data began in earnest and lasted for about a year and a half before we had sufficient data from which we could draw an initial conclusion.

As we said earlier, we expected solar neutrino events in the modified Kamioka detector at the agonizingly slow rate of one event every two or three days—if the detector was working as planned and if the measurement made by the radiochemical experiment in the Homestake Mine was correct. You may wonder what the hoped-for goal was of all the hard work and money spent on these modifications. One answer is the elusive nature of the neutrino. The Sun provides an abundant supply of low-energy neutrinos that we believe to be of well-known origin and that had to travel through dense solar matter and high solar magnetic fields and finally travel from the Sun to the Earth before we could detect them in the Kamioka mine. One can imagine many adventures befalling them on that voyage, adventures that might reduce their number, change one of their intrinsic properties, or even create as yet unknown particle products. Any of these might reveal an unexpected aspect of neutrinos and thereby influence our view not only of neutrinos but also of the extended family of elementary particles.

Another answer to what goals we were seeking was given in Chapter 3, which explained that neutrinos from the Sun serve as a probe of its interior and consequently provide a means of testing the validity of our models of the Sun. Astronomers, notably John Bahcall of the Institute for Advanced Study in Princeton, have sharpened these models over several decades, so by now they offer a remarkably keen mathematical edge to our understanding of the Sun.

At the average rate of one event every two or three days, we needed a data-collection period of at least a year to obtain a modest but statistically significant sample of solar neutrino events. Fortunately, there was much to keep us occupied. The detector and its auxiliary apparatus had to be maintained and its performance monitored, and the data collected and analyzed. To perform the required analysis, we had to devise computer programs with a greater capability than those used to analyze the supernova neutrino data. In particular, we needed to discriminate between background events from the radioactivity in the detector water and background events induced in the detector by cosmic rays. These demanded greater skill and sophistication in the analysis than what we were using to detect the antineutrino signal from SN 1987A, a signal almost entirely free of background events (see Color Plate 9).

To ensure that we were doing this efficiently and were correctly finding the solar signal, we organized two largely independent analysis efforts—one in Tokyo and one in Philadelphia—each of which started with copies of the magnetic tapes holding the raw data and then proceeded independently. Imagine our euphoria when we found that the two sets of results, obtained after extensive calculations and months of study, agreed in detail with only minor differences! (To digress, in modern physics and astronomy, which use elaborate computer programs and high-speed computers to process data with a quantity of detail never before possible, the need for independent analyses with many checks for consistency has become an absolute necessity.)

As we noted earlier, the purpose of the Kamioka solar neutrino experiment was to demonstrate that the electrons observed by the detector were in fact produced by neutrinos originating in the Sun. We knew that the direction of motion of the electrons that were hit by the solar neutrinos would closely mirror the direction of the neutrinos, and we relied on this property to correlate the observed electrons with the neutrinos from the Sun. It was also this property that had to be the principal identifying signature of the solar neutrino signal we wanted to find. To use it, we needed to know the continuously changing position of the Kamioka detector with respect to the Sun as the Earth rotated about its own axis and circled the Sun. It was imperative that this feature be taken into account and be incorporated in our data analysis programs.

Our method of detection and the directional property of the neutrino–electron reaction were not called into question because they had been used successfully in laboratory experiments, but our segment of the scientific community was skeptical that we would be able to use them with equal success in a real detector as large as the one at Kamioka. The specific question they raised was whether we would be able to point the observed events back to the Sun with sufficient accuracy over such a long distance. We, too, suffered similar misgivings and occasional losses of confidence during the year it took us to understand and overcome the background events that initially plagued the experiment.

After 450 days of acquiring data, we obtained the preliminary result shown in Figure 8 to illustrate the success of the method. The directions of the electrons observed in the detector could in fact be used to distinguish the background events from the electrons constituting the signal from the Sun and to point the neutrinos producing those electrons back to the Sun. At the right side of Figure 8, where the angle between the direction of an observed electron and a line from the

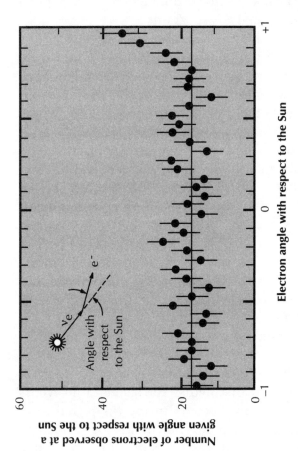

FIGURE 8. Plot showing the angular correlation of the earliest solar neutrino events in the Kamioka detector with the direction from the Sun. The peak on the right-hand side of the plot indicates solar neutrino events. The other data points are not correlated with the direction of the Sun.

Sun to the detector is small, there is a clear excess of observed electrons, indicating that the parent neutrinos emerged from the Sun. Elsewhere in the plot of Figure 8, the observed electrons constitute an irreducible background having no angular correlation with the Sun, which we were unable to eliminate completely. To make the plot in Figure 8, we needed a computer program to predict the location of the detector relative to the Sun at any instant of the day or night. As we saw in the case of the antineutrinos from SN 1987A, the Earth is not a barrier to antineutrinos or neutrinos, and solar neutrinos reached the Kamioka detector at night as well as during the day. The plot in Figure 8 thus includes electrons observed at all times, day and night.

When we published the measured results in Figure 8, the skepticism concerning the method and our ability to implement it soon disappeared. With significantly more solar neutrino data, obtained later at Kamioka, the scientific community dispensed with its remaining doubts and came to accept the measured yields of solar neutrinos by the Homestake and Kamioka detectors. Furthermore, from Kamioka's real-time, directional data, we could conclude that the events observed in the two detectors were, without serious doubt, initiated by neutrinos from the Sun and that these could only be the product of fusion reactions in the Sun.

There is much more to the story of solar neutrinos. The agreement between the solar model and the measurement in Figure 8 is not perfect, nor is the agreement between the Homestake and Kamioka measurements. In addition, when the results from new radiochemical detectors in Italy and Russia, which use the rare element gallium as detector material, were compared with those from Homestake and Kamioka, we were led to suspect that solar neutrinos may in fact undergo an adventure on their voyage to Earth. Considered in their entirety, the data seem to show that the neutrino mass may not be zero and that it may be measured by further study of solar neutrinos. This prospect led to the construction of three more underground solar neutrino detectors, one a much larger companion of the original Kamioka detector located in the same mine in Japan and two with special properties, one in a very deep mine in Sudbury, Canada, and one in the same laboratory in the Gran Sasso tunnel in Italy as the gallium experiment. Results from these new detectors in combination with those from the older detectors may well provide in the near future a conclusive answer to the problem of neutrino mass.

APPENDIX 1.
POWERS OF 10
NOTATION

We offer here an easy-to-learn, easy-to-use numerical shorthand that is commonly used in physics and astronomy because it is simple and concise. You may want to spend a few minutes to master the shorthand—known as the *powers of 10*, or *scientific notation*—to help you grasp and appreciate the extremely large and extremely small quantities encountered in this book.

First, look at an example that is easily generalized. The number 1,000,000 can be written as $10 \times 10 \times 10 \times 10 \times 10 \times 10$, that is, as the result of multiplying 10 by itself 6 times. In the powers of 10 notation, this is written $10^1 \times 10^1 \times 10^1 \times 10^1 \times 10^1 \times 10^1$, which can be condensed by writing 10^6 in place of the cumbersome $10^1 \times 10^1 \times 10^1 \times 10^1 \times 10^1 \times 10^1$. The number 6 is simply the sum of the raised numbers—called *exponents*—standing to the right of each factor of 10 in $10^1 \times 10^1 \ldots 10^1$, and it tells us that 6 products or powers of 10 are required to denote 1 million. Similarly, 10^9, or 9 powers of 10, denote 1 billion.

The preceding paragraph is a definition of multiplication. The only other numerical operation necessary for our purpose is division. For example, 100 divided by 10 (usually written as 100/10) may also be written in the powers of 10 notation as $10^2/10^1 = 10^1$, which suggests, and in fact correctly indicates, that division is carried out by subtracting the exponents, just as multiplication is done by adding the exponents. In another example, a billion divided by a million is $10^9/10^6 = 10^3$, telling us that a billion is one thousand times larger than a million.

The notation is consistent and completely clear once you recognize that $10^0 = 10^1/10^1 = 10^2/10^2$, and so on, is just equal to 1. Then we can write the powers of 10 as $10^0 = 1, 10^1 = 10, 10^2 = 100, 10^3 = 1000, \ldots 10^6 = 1,000,000$, which is where we began.

To express small numbers, that is, numbers less than 1, we can write the number as a fraction with 10^0 in the numerator and a larger power of 10 in the denominator. Examples are $0.1 =$

$10^0/10^1$ (= 1/10), and 0.001 = $10^0/10^3$ (= 1/1000). This can be simplified even further by applying the rule for division, that is, subtracting the exponents. Then 0.1 = 10^{-1}, and 0.001 = 10^{-3}, and in general, a number less than 1 is expressed by a negative exponent.

Finally, any power of 10 may be multiplied or divided by any real number, for example, 170,000 = 1.70×10^5, or 20 = 2×10^1 = 0.2×10^2, and so forth. Hence, all real numbers can be written in the powers of 10 notation.

You may want to practice using some examples you can look up in an encyclopedia or a physics textbook, for instance, the radius of a hydrogen atom in centimeters, the radius of the Earth in meters, and the distance from the Sun to the Earth in meters. If you can express these figures in the shorthand, you will have mastered the powers of 10.

APPENDIX 2.
UNITS OF ENERGY

In our technologically advanced societies, we usually measure a quantity of energy in terms of power and time, for example, as so many kilowatts for so many hours. This is a convenient way for a consumer to calculate the cost of power usage, and it is technically correct. But it is cumbersome for most purposes in physics and astronomy, in which it is more efficient to refer to a unit of energy directly.

Many physics phenomena involve electrically charged particles that are acted on by an electric force. As a consequence, a frequently used unit of energy is that amount acquired by a charged particle, or the natural fundamental unit of charge when it is acted on by the electric force due to a voltage difference of, say, 1 volt from a small cell. The fundamental unit of electric charge in nature is the charge on the electron, which is exactly equal (apart from sign) to the charge on the proton—otherwise, neutral atoms would be rare indeed. It may be helpful to think of an electron that is free to move inside a wire, the ends of which are attached to the terminals of a 1-volt cell. In moving from one end of the wire to the other, the electron gains a quantity of energy equal to the magnitude of its charge times 1 volt. If the electron charge is expressed in terms of the coulomb (a unit of charge named after an early-nineteenth-century French physicist), the quantity of charge carried by an electron is 1.6×10^{-19} coulomb. When accelerated by a 1-volt battery, the resultant energy of the electron is 1.6×10^{-19} coulomb $\times$ 1 volt $= 1.6 \times 10^{-19}$ joule (which, you guessed it, was named for another physicist, this time an Englishman of the late nineteenth century).

For much usage the joule also is too cumbersome, and so a briefer terminology has been adopted. Instead of carrying around the number 1.6×10^{-19} coulomb as the electron charge, we simply refer to it as an *electron*, where the word *electron* stands for the charge value 1.6×10^{-19}

coulomb. Then the energy acquired by the electron in the preceding example may be expressed as 1 "electron" × 1 volt, which is read as 1 electron-volt and is customarily abbreviated as 1 eV. Since the energy of the electron must be the same in whatever units it is expressed, 1 eV must equal 1.6×10^{-19} joule, a result that, mysteriously, has always plagued students of elementary physics.

If a particle carrying twice the charge of the electron were to be accelerated by a 1-volt cell, its energy would be 2 "electron" × 1 volt, or 2 eV, and if it were accelerated by a 2-volt cell, the final energy would be 4 eV. A thousand eV is a kilo-eV, or keV, and a million eV is a mega-eV, or MeV. These are the convenient units of energy in atomic, nuclear, and particle physics used by physicists and astronomers. But don't confuse the quantities in that submicroscopic world with those in our everyday macroscopic world. For example, one MeV corresponds to an energy of 1.6×10^{-13} joule, and a 100-watt lightbulb uses energy at the rate of 100 joules every second it is lit. In general, the two worlds are many orders of magnitude apart in energy.

APPENDIX 3.
THE PERIODIC
TABLE OF THE
CHEMICAL
ELEMENTS

Appendix 3 shows the Periodic Table of the Chemical Elements, which can also be found in any elementary textbook of chemistry or modern physics. The numbers running along the left edge of the table specify the periods, and the roman numbers at the top delineate the chemical groups. For example, Group I contains the chemical elements known as the alkali atoms—reactive elements such as lithium (Li) and sodium (Na), with a single atomic electron outside a closed electron shell—plus hydrogen (H). Another example is Group VIII, composed of the noble (nonreactive) gases, helium (He), neon (Ne), through radon (Ra), which have closed electron shells.

	I	II	III	IV	V	VI	VII	VIII		
1	1 H 1.0080							2 He 4.003		
2	3 Li 6.940	4 Be 9.03	5 B 10.82	6 C 12.010	7 N 14.008	8 O 16.0000	9 F 19.00	10 Ne 20.183		
3	11 Na 22.997	12 Mg 24.32	13 Al 26.98	14 Si 28.09	15 P 30.975	16 S 32.066	17 Cl 35.457	18 Ar 39.944		
4	19 K 39.100	20 Ca 40.08	21 Sc 44.96	22 Ti 47.90	23 V 50.95	24 Cr 52.01	25 Mn 54.93	26 Fe 55.85	27 Co 58.94	28 Ni 58.69
	29 Cu 63.54	30 Zn 65.38	31 Ga 69.72	32 Ge 72.60	33 As 74.91	34 Se 78.96	35 Br 79.916	36 Kr 83.80		
5	37 Rb 85.48	38 Sr 87.63	39 Y 88.92	40 Zr 91.22	41 Nb 92.91	42 Mo 95.95	43 Tc [99]	44 Ru 101.7	45 Rh 102.91	46 Pd 106.7
	47 Ag 107.880	48 Cd 112.41	49 In 114.76	50 Sn 118.70	51 Sb 121.76	52 Te 127.61	53 I 126.91	54 Xe 131.3		
6	55 Cs 132.91	56 Ba 137.36	57–71 Rare Earths[a]	72 Hf 178.6	73 Ta 180.88	74 W 183.92	75 Re 186.31	76 Os 190.2	77 Ir 193.1	78 Pt 195.23
	79 Au 197.2	80 Hg 200.61	81 Tl 204.39	82 Pb 207.21	83 Bi 209.00	84 Po 210	85 At [210]	86 Rn 222		
7	87 Fr [223]	88 Ra 226.05	89–96 Actinide Series[b]							

a *Rare Earth or Lanthanide Series*

57	La 138.92	58	Ce 140.13	59	Pr 140.92	60	Nd 144.27	61	Pm [147]
62	Sm 150.43	63	Eu 152.0	64	Gd 156.9	65	Tb 159.2	66	Dy 162.46
67	Ho 164.94	68	Er 167.2	69	Tm 169.4	70	Yb 173.04	71	Lu 174.99

b *Actinide Series*

89	Ac 227	90	Th 232.12	91	Pa 231	92	U 238.07
93	Np [237]	94	Pu [242]	95	Am [243]	96	Cm [243]
97	Bk [245]	98	Cf [246]				

The number next to the element symbol is the atomic number, that is, the number of protons in the nucleus and, equivalently, the total number of electrons in the electron shells. The number below the element symbol is the atomic mass (in grams) of 1 mole of the element. Less than 1 percent of the atomic mass is occupied by the atomic electrons. Uranium (U) is found in the actinide series.

Several elements in Periods 1 and 2 were formed by fusion in main-sequence stars. The remainder of the elements, as far as we know, were produced by supernovae.

BIBLIOGRAPHY

The following is a short list of articles and books, not all on supernovae, that might profitably be scanned or even given more time if they kindle interest.

W. David Arnett, *Supernovae and Nucleosynthesis: An Investigation of the History of Matter, from the Big Bang to the Present,* Princeton University Press, Princeton, NJ (1996).

W. David Arnett, John N. Bahcall, Robert P. Kirshner, and Stanford E. Woosley, "Supernova 1987A," *Annual Review of Astronomy and Astrophysics,* **27,** 629 (1989).

Walter Baade and Fritz Zwicky, "Supernovae and Cosmic Rays," *The Physical Review,* **45,** 138 (1934).

John N. Bahcall, *Neutrino Astrophysics*, Cambridge University Press, Cambridge (1989).

Jeremy Bernstein, *Experiencing Science*, Basic Books, New York (1978).

D. D. Clayton, *Principles of Stellar Evolution and Nucleosynthesis*, University of Chicago Press, Chicago (1983).

I. J. Danziger and K. Kjär, eds., *SN 1987A and Other Supernovae*, ESO Workshop and Conference Proceedings, no. 37 (1991).

Albert Einstein, *Ideas & Opinions*, trans. S. Bargmann, Dell, New York (1978).

G. Ellis, A. Lanza, and J. Miller, eds., *Renaissance of General Relativity and Cosmology*, Cambridge University Press, Cambridge (1993).

L. M. Krauss and S. Tremaine, "Test of the Weak Equivalence Principle for Neutrinos and Photons," *Physical Review Letters*, **60,** 176 (1988).

M. J. Longo, "New Precision Tests of the Einstein Equivalence Principle from SN 1987A," *Physical Review Letters*, **60,** 173 (1988).

Laurence A. Marschall, *The Supernova Story*, Plenum Press, New York (1988).

Philip Morrison, *Powers of Ten*, Scientific American Library, New York (1982).

Paul Murdin, *End in Fire: The Supernova in the Large Magellanic Cloud*, Cambridge University Press, Cambridge (1990).

Georg G. Raffelt, *Stars as Laboratories for Fundamental Physics*, University of Chicago Press, Chicago (1996).

Carl Sagan, *Shadows of Forgotten Ancestors*, Random House, New York (1992).

D. W. Sciama, *The Unity of the Universe*, Anchor Books, Doubleday, New York (1961).

Stuart L. Shapiro and Saul A. Teukolsky, *Black Holes, White Dwarfs, and Neutron Stars*, Wiley, New York (1983).

Lewis Thomas, *The Youngest Science*, Viking Press, New York (1983).

Barbara Ward, *Spaceship Earth*, Columbia University Press, New York (1966).

James D. Watson, *The Double Helix: A Personal Account of the Discovery of the Structure of DNA*, Norton, New York (1980).

Kurt W. Weiler and Richard A. Sramek, "Supernovae and Supernova Remnants," *Annual Review of Astronomy and Astrophysics*, **26,** 295 (1988).

Steven Weinberg, *The First Three Minutes: A Modern View of the Origin of the Universe*, Basic Books, New York (1977).

There also are many entries on the Web relating to Supernova 1987A. Any search program, such as Alta Vista (query Supernova + 1987A), will turn up a multitude of them. For example, *http://reudi.mit.edu/AstroDay/ sn1987.html* (4 May 95) discusses new insights in astronomy from SN 1987A, and *http://lheawww.gsfc.nasa.gov/docs/xray/ staff//tyler/hubble/sn1987a.html* (20 June 96) describes in more detail the ring structure shown in Figure 7 of the text.

GLOSSARY

Binding energy The energy released in the formation of a bound system from an assembly of independent objects. In particular, the binding energy of an atomic nucleus is the energy released in the formation of the nucleus. Correspondingly, it is the energy required to break up a bound system.

Black hole An object of very small dimensions and such high density that neither light nor neutrinos can escape from it.

Decay For any radioactive element or particle, the length of time, often specified quantitatively as its half-life, for a fraction (one-half) of the atoms or particles to disappear.

Electromagnetic radiation Electromagnetic particles or waves carrying energy and momentum. Various types of electromagnetic radiation include gamma rays, X rays, ultraviolet, visible, infrared, and radio waves, in order of decreasing energy.

Electron A negatively charged particle with a charge precisely equal in magnitude to the positive charge on the proton. Electrons are bound in atoms by the electrical force between the nucleus and the electrons, and they are largely responsible for the chemical properties of different atoms.

Element A basic chemical entity specified by the number of protons in its nucleus (atomic number) and the equal number of bound electrons outside the nucleus. See the Periodic Table of the Elements in Appendix 3.

Elementary particle One of what are thought now to be the 12 fundamental building blocks of all matter. These are elementary in the sense that they are observed to have no internal structure of any kind. In addition, 13 other elementary objects serve to transmit the fundamental forces between particles.

Fission The breakup, caused by the addition of another neutron, of a heavy nucleus (such as uranium) into two lighter nuclei and several neutrons with the emission of about 200 MeV of energy.

Fusion The prototype fusion reaction is the congealing of four protons into a helium nucleus with the emission of two positively charged electrons, two neutrinos, and 29 MeV of energy. Other elements lighter than iron also can fuse to heavier nuclei with the emission of energy.

Galaxy A large (10^8 to 10^{13} solar masses), gravitationally bound aggregate of stars and interstellar matter.

Large Magellanic Cloud One of the nearest galaxies to the Milky Way, about 160,000 light-years away from the Earth.

Luminosity The rate of energy emission from a source; for example, the Sun emits 10^{26} joules per second (see Appendix 2).

Magnetic field A region of space where electrically charged particles are observed to deviate from linear motion and occasionally accelerated. Magnetic fields exist on the Earth, on stars, and in interstellar space.

Main-sequence star A star that is burning hydrogen in its core. Ninety percent of the stars we observe are main-sequence stars.

Milky Way A band of light composed of many stars and nebulae lying near the galactic plane. It is the name of the galaxy in which our solar system resides.

Neutrino An elementary particle, presumably stable against decay, with no electric charge and no as yet measurable rest mass, that carries energy and momentum. Its main characteristic is the weakness of its interactions with all other particles.

Neutron One of the constituents of nuclei with zero electrical charge and a mass slightly greater than that of the proton, the other nuclear constituent. The free (that is, unbound) neutron at rest decays with a half-life of about 1000 seconds into a proton, an electron, and an antineutrino.

Neutron star A very dense star composed primarily of neutrons, roughly 10 to 20 kilometers in diameter with a very high magnetic field. Pulsars are thought to be rotating magnetic neutron stars.

Nuclear constituent See neutron and proton.

Nucleosynthesis The series of nuclear fusion reactions leading to the quantitative explanation of the abundance of the lighter nuclei—deuterium, helium, and lithium—in agreement with observation. The synthesis of heavier nuclei is through supernovae.

Nucleus The massive, positively charged central core of an atom around which the atomic electrons circulate. The radii of nuclei lie in the range 10^{-13} to 10^{-12} cm, and the radii of atoms are 10,000 to 100,000 times larger. Almost all the mass of an atom is in the nucleus, which has a density (mass per volume) of roughly 10^{14} gm per cubic centimeter.

Photon A fundamental particle/wave of light carrying discrete quantities of energy and momentum. Photons have zero rest mass and are their own antiparticles.

Photosensitive element A glass envelope enclosing a very thin metallic layer on one of its inner surfaces that emits electrons when struck by photons. The initially emitted electrons are multiplied in the remainder of the element to give measurable electric signals. Often called a *photomultiplier tube*.

Proton A positively charged fundamental particle that is one of the constituents of nuclei; the nucleus of the hydrogen atom.

Pulsar A stellar object that emits pulses of radio waves with great regularity; believed to be a rapidly rotating, magnetized neutron star.

Radioactivity The spontaneous emission of particles or gamma rays by an unstable nucleus.

Radio telescope A large antenna capable of receiving radio-band signals from radio sources outside the solar system.

Sanduleak 202, or Sk-69-202 The blue supergiant star in the Large Magellanic Cloud that became SN 1987A.

Schwarzschild radius The radius at which a spherically symmetric body of a given mass becomes a black hole, according to the general theory of relativity.

Shock A sharp change in the pressure, temperature, and density of a fluid substance that occurs when the velocity of the fluid exceeds the velocity of sound in the fluid.

Solar mass The mass of the Sun equal to approximately 2×10^{30} kilograms.

Special theory of relativity The theory providing the quantitative understanding of measurements made when the observer and the event are in motion with constant velocity of one relative to the other. The general theory of relativity generalizes Newton's law of gravitation for masses accelerated by their mutual attraction.

Supernova remnant The gaseous nebula and the presumed neutron star at its center following a supernova explosion.

Universe The total cosmos: in current cosmology, if the universe is bounded, the mass is of the order of 10^{56} gm, radius 2×10^{28} cm, with an age between 10 and 20×10^9 yr.

INDEX